D1329402

For

Madge Goldman

who has done so much
to bring so many people together

I.M. Gelfand
Mark Saul

Trigonometry

Birkhäuser
Boston • Basel • Berlin

I.M. Gelfand
Department of Mathematics
Rutgers University
Piscataway, NJ 08854-8019
U.S.A.

Mark Saul
Bronxville Schools
Bronxville, NY 10708
U.S.A.

Library of Congress Cataloging-in-Publication Data

Gel'fand, I.M. (Izrail' Moiseevich)
Trigonometry / I.M. Gel'fand, Mark Saul.
p. cm.
ISBN 0-8176-3914-4 (alk. paper)
1. Trigonometry, Plane. I. Saul, Mark E. II. Title.
QA533.G45 1999
516.24'2–dc21 99-32245
CIP

Printed on acid-free paper

Birkhäuser 

ISBN 0-8176-3914-4 SPIN 10536370
ISBN 3-7643-3914-4

Typeset in LaTeX by Martin Stock, Cambridge, MA
Printed and bound by Hamilton Printing, Rensselaer, NY
Printed in the United States of America

9 8 7 6 5 4 3 2 1

Contents

Preface

In a sense, trigonometry sits at the center of high school mathematics. It originates in the study of geometry when we investigate the ratios of sides in similar right triangles, or when we look at the relationship between a chord of a circle and its arc. It leads to a much deeper study of periodic functions, and of the so-called *transcendental* functions, which cannot be described using finite algebraic processes. It also has many applications to physics, astronomy, and other branches of science.

It is a very old subject. Many of the geometric results that we now state in trigonometric terms were given a purely geometric exposition by Euclid. Ptolemy, an early astronomer, began to go beyond Euclid, using the geometry of the time to construct what we now call tables of values of trigonometric functions.

Trigonometry is an important introduction to calculus, where one studies what mathematicians call *analytic* properties of functions. One of the goals of this book is to prepare you for a course in calculus by directing your attention away from particular values of a function to a study of the function as an object in itself. This way of thinking is useful not just in calculus, but in many mathematical situations. So trigonometry is a part of pre-calculus, and is related to other pre-calculus topics, such as exponential and logarithmic functions, and complex numbers. The interaction of these topics with trigonometry opens a whole new landscape of mathematical results. But each of these results is also important in its own right, without being "pre-" anything.

We have tried to explain the beautiful results of trigonometry as simply and systematically as possible. In many cases we have found that simple problems have connections with profound and advanced ideas. Sometimes we have indicated these connections. In other cases we have left them for you to discover as you learn more about mathematics.

About the exercises: We have tried to include a few problems of each "routine" type. If you need to work more such problems, they are easy to find. Most of our problems, however, are more challenging, or exhibit a new aspect of the technique or object under discussion. We have tried to make each exercise tell a little story about the mathematics, and have the stories build to a deep understanding.

We will be happy if you enjoy this book and learn something from it. We enjoyed writing it, and learned a lot too.

Acknowledgments

The authors would like to thank Martin Stock, who took a very ragged manuscript and turned it into the book you are now holding. We also thank Santiago Samanca and the late Anneli Lax for their reading of the manuscript and correction of several bad blunders. We thank Ann Kostant for her encouragement, support, and gift for organization. We thank the students of Bronxville High School for their valuable classroom feedback. Finally, we thank Richard Askey for his multiple readings of the manuscripts, for correcting some embarrassing gaffes, and for making important suggestions that contributed significantly to the content of the book.

Israel M. Gelfand
Mark Saul
March 20, 2001

Chapter 0

Trigonometry

f. Gr. $\tau\rho\acute{\iota}\gamma\omega\nu o$ - ν triangle + - $\mu\epsilon\tau\rho\acute{\iota}\alpha$ measurement.
— *Oxford English Dictionary*

In this chapter we will look at some results in geometry that set the stage for a study of trigonometry.

1 What is new about trigonometry?

Two of the most basic figures studied in geometry are the triangle and the circle. Trigonometry will tell us more than we learned in geometry about each of these figures.

For example, in geometry we learn that if we know the lengths of the three sides of a triangle, then the measures of its angles are completely determined[1] (and, in fact, almost everything else about the triangle is determined). But, except for a few very special triangles, geometry does not tell us how to compute the measures of the angles, given the measures of the sides.

Example 1 The measures of the sides of a triangle are 6, 6, and 6 centimeters. What are the measures of its angles?

[1] It is sometimes said that the lengths of three sides determine a triangle, but one must be careful in thinking this way. Given three arbitrary lengths, one may or may not be able to form a triangle (they form a triangle if and only if the sum of any two of them is greater than the third). But if one can form a triangle, then the angles of that triangle are indeed determined.

Solution. The triangle has three equal sides, so its three angles are also equal. Since the sum of the angles is 180°, the degree-measure of each angle is 180/3 = 60°. Geometry allows us to know this without actually measuring the angles, or even drawing the triangle. □

Example 2 The measures of the sides of a triangle are 5, 6, and 7 centimeters. What are the measures of its angles?

Solution. We cannot find these angle measures using geometry. The best we can do is to draw the triangle, and measure the angles with a protractor. But how will we know how accurately we have measured? We will answer this question in Chapter 3. □

Example 3 Two sides of a triangle have length 3 and 4 centimeters, and the angle between them is 90°. What are the measures of the third side, and of the other two angles?

Solution. Geometry tells us that if we know two sides and an included angle of a triangle, then we ought to be able to find the rest of its measurements. In this case, we can use the Pythagorean Theorem (see page 7) to tell us that the third side of the triangle has measure 5. But geometry will not tell us the measures of the angles. We will learn how to find them in Chapter 2. □

Exercise Using a protractor, measure the angles of the triangle below as accurately as you can. Do your measurements add up to 180°?

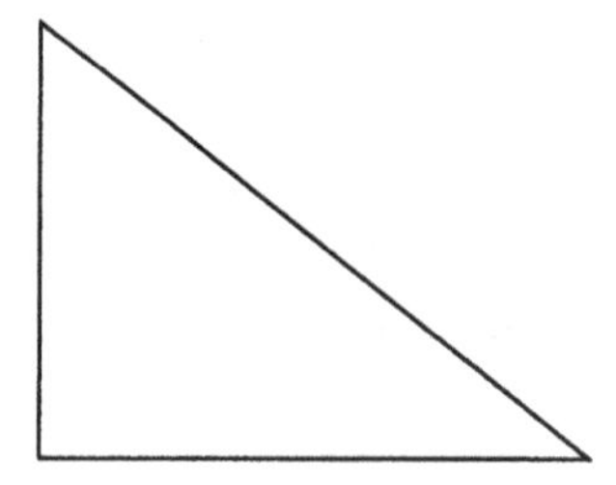

Let us now turn our attention to circles.

Example 4 In a certain circle, a central angle of 20° cuts off an arc that is 5 inches long. In the same circle, how long is the arc cut off by a central angle of 40°?

Solution. We can divide the 40°angle into two angles of 20°. Each of these angles cuts off an arc of length 5″, so the arc cut off by the 40°angle is $5 + 5 = 10$ inches long.

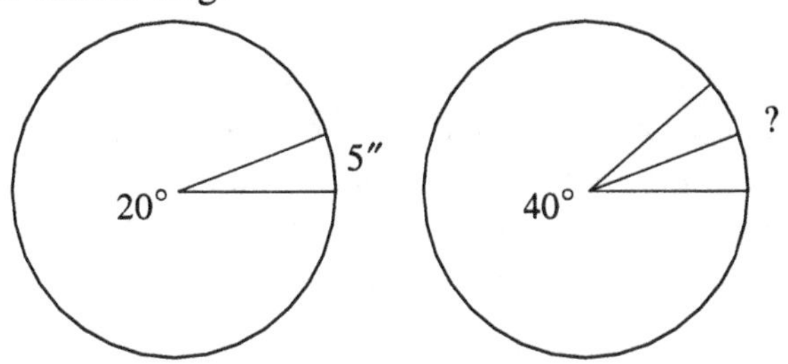

That is, if we double the central angle, we also double the length of the arc it intercepts. □

Example 5 In a certain circle, a central angle of 20° determines a chord that is 7 inches long. In the same circle, how long is the chord determined by a central angle of 40°?

Solution. As with Example 4, we can try to divide the 40° angle into two 20° angles:

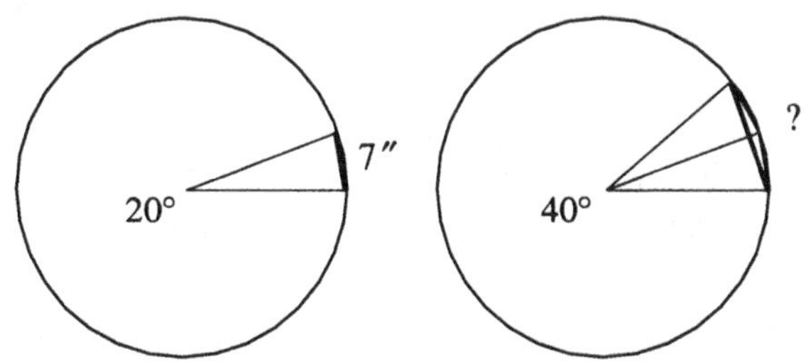

However, it is not so easy to relate the length of the chord determined by the 40° angle to the lengths of the chords of the 20° angles. Having doubled the angle, we certainly have not doubled the chord. □

Exercises

1. In a circle, suppose we draw any central angle at all, then draw a second central angle which is larger than the first. Will the arc of the second central angle always be longer than the arc of the first? Will the chord of the second central angle also be larger than the chord of the first?

2. What theorem in geometry guarantees us that the chord of a 40° angle is less than double the chord of a 20° angle?

3. Suppose we draw any central angle, then double it. Will the chord of the double angle always be less than twice the chord of the original central angle?

Trigonometry and geometry tell us that any two equal arcs in the same circle have equal chords; that is, if we know the measurement of the arc, then the length of the chord is determined. But, except in special circumstances, geometry does not give us enough tools to calculate the length of the chord knowing the measure of the arc.

Example 6 In a circle of radius 7, how long is the chord of an arc of 90°?

Solution. If we draw radii to the endpoints of the chord we need, we will have an isosceles right triangle:

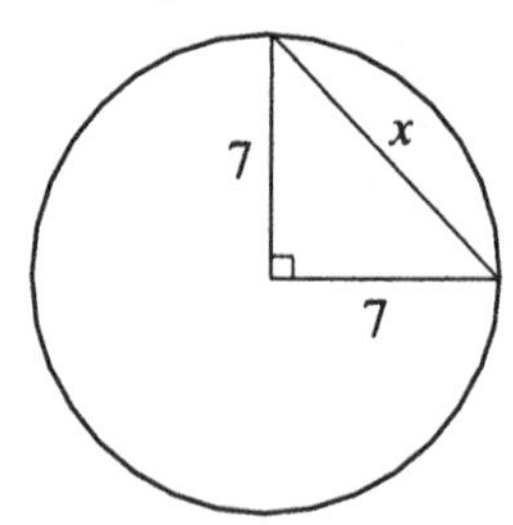

Then we can use the Pythagorean Theorem to find the length of the chord. If this length is x, then $7^2 + 7^2 = x^2$, so that $x = \sqrt{98} = 7\sqrt{2}$. □

Example 7 In a circle of radius 7, how long is the chord of an arc of 38°?

Solution. Geometry does not give us the tools to solve this problem. We can draw a triangle, as we did in Example 6:

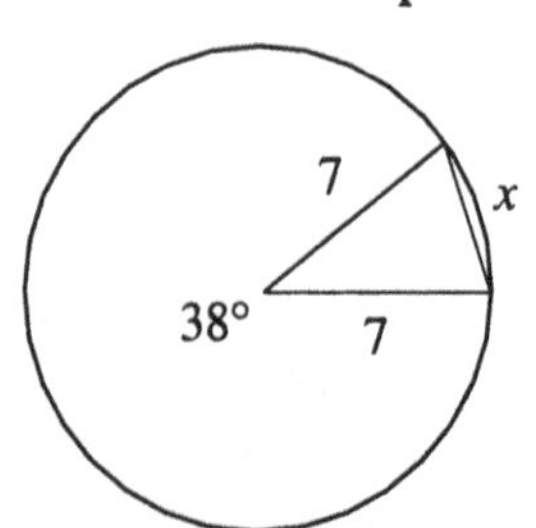

But we cannot find the third side of this triangle using only geometry. However, this example does illustrate the close connection between measurements in a triangle and measurements in a circle. □

Exercises

1. What theorem from geometry guarantees us that the triangle in the diagram for Example 7 is completely determined?

2. Note that the triangle in Example 7 is isosceles. Calculate the measure of the two missing angles.

Trigonometry will help us solve all these kinds of problems. However, trigonometry is more than just an extension of geometry. Applications of trigonometry abound in many branches of science.

Example 8 Look at any pendulum as it swings. If you look closely, you will see that the weight travels very slowly at either end of its path, and picks up speed as it gets towards the middle. It travels fastest during the middle of its journey. □

Example 9 The graph below shows the time of sunrise (corrected for daylight savings) at a certain latitude for Wednesdays in the year 1995. The data points have been joined by a smooth curve to make a continuous graph over the entire year.

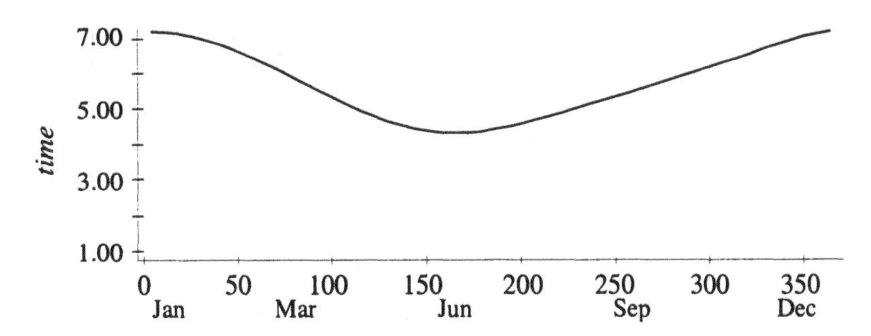

We expect this curve to be essentially the same year after year. However, neither geometry nor algebra can give us a formula for this curve. In Chapter 8 we will show how trigonometry allows us to describe it mathematically. Trigonometry allows us to investigate any periodic phenomenon – any physical motion or change that repeats itself. □

2 Right triangles

We will start our study of trigonometry with triangles, and for a while we will consider only right triangles. Once we have understood right triangles, we will know a lot about other triangles as well.

Suppose you wanted to use e-mail to describe a triangle to your friend in another city. You know from geometry that this usually requires three pieces of information (three sides; two sides and the included angle; and so on). For a right triangle, we need only two pieces of information, since we already know that one angle measures 90°.

In choosing our two pieces of information, we must include at least one side, so there are four cases to discuss:

a) the lengths of the two legs;

b) the lengths of one leg and the hypotenuse;

c) the length of one leg and the measure of one acute angle;

d) the length of the hypotenuse and the measure of one acute angle.

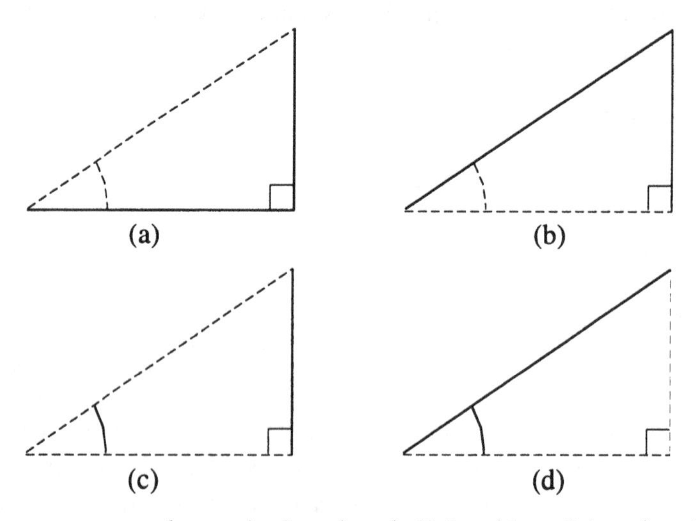

Suppose we want to know the lengths of all the sides of the triangle. For cases (a) and (b) we need only algebra and geometry. For cases (c) and (d), however, algebraic expressions do not (usually) suffice. These cases will introduce us to trigonometry, in Chapter 1.

3 The Pythagorean theorem

We look first at the chief geometric tool which allows us to solve cases (a) and (b) above. This tool is the famous *Pythagorean Theorem*. We can separate the Pythagorean theorem into two statements:

Statement I: If a and b are the lengths of the legs of a right triangle, and c is the length of its hypotenuse, then $a^2 + b^2 = c^2$.

Statement II: If the positive numbers a, b, and c satisfy $a^2 + b^2 = c^2$, then a triangle with these side lengths has a right angle opposite the side with length c.[2]

These two statements are *converses* of each other. They look similar, but a careful reading will show that they say completely different things about triangles. In the first statement, we know something about an angle of a triangle (that it is a right angle) and can conclude that a certain relationship holds among the sides. In the second statement, we know something about the sides of the triangle, and conclude something about the angles (that one of them is a right angle).

The Pythagorean theorem will allow us to reconstruct a triangle, given two legs or a leg and the hypotenuse. This is because we can find, using this information, the lengths of all three sides of the triangle. As we know from geometry, this completely determines the triangle.

Example 10 In the English university town of Oxford, there are sometimes lawns occupying rectangular lots near the intersection of two roads (see diagram).

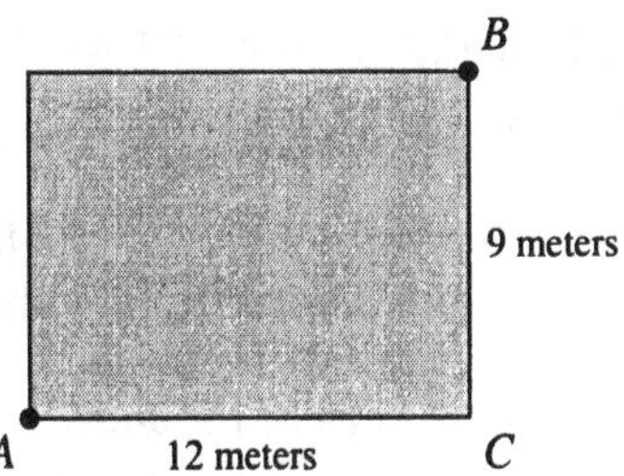

[2]In fact, we can make a stronger statement than statement II:

Statement II′: If the positive numbers a, b, and c satisfy $a^2 + b^2 = c^2$, then there exists a triangle with sides a, b, and c, and this triangle has a right angle opposite the side with length c.

This statement includes, for example, the fact that if $a^2 + b^2 = c^2$, then $a + b > c$.

In such cases, professors (as well as small animals) are allowed to cut across the lawn, while students must walk around it. If the dimensions of the lawn are as shown in the diagram, how much further must the students walk than the professors in going from point A to point B?

Solution. Triangle ABC is a right triangle, so statement I of the Pythagorean theorem applies:

$$\begin{aligned} AB^2 &= AC^2 + BC^2 \\ &= 12^2 + 9^2 \\ &= 144 + 81 = 225\,. \end{aligned}$$

So $AB = 15$ meters, which is how far the professor walks.

On the other hand, the students must walk the distance $AC + CB = 12 + 9 = 21$. This is 6 meters longer than the professor's walk, or 40% longer. □

Example 11 Show that a triangle with sides 3, 4, and 5 is a right triangle.

Solution. We can apply statement II to see if it is a right triangle. In fact, $5^2 = 25 = 3^2 + 4^2$, so the angle opposite the side of length 5 is a right angle. Notice that we cannot use statement I of the Pythagorean theorem to solve this problem. □

Exercises The following exercises concern the Pythagorean theorem. In solving each problem, be sure you understand which of the two statements of this theorem you are using.

1. Two legs of a right triangle measure 10 and 24 units. Find the length of the hypotenuse in the same units.

2. The hypotenuse of a right triangle has length 41 units, and one leg measures 9 units. Find the measure of the other leg.

3. Show that a triangle with sides 5, 12, and 13 is a right triangle.

4. One leg of a right triangle has length 1 unit, and the hypotenuse has length 3 units. What is the length of the other leg of the triangle?

5. The hypotenuse of an isosceles right triangle has length 1. Find the length of one of the legs of this triangle.

6. In a right triangle with a 30°angle, the hypotenuse has length 1. Find the lengths of the other two legs.

   Hint: Look at the diagram in the footnote on page 11.

7. Two points, A and B, are given in the plane. Describe the set of points X such that $AX^2 + BX^2 = AB^2$.

   (Answer: A circle with its center at the midpoint of AB.)

8. Two points, A and B, are given in the plane. Describe the set of points for which $AX^2 - BX^2$ is constant.

4 Our best friends (among right triangles)

There are a few right triangles which have a very pleasant property: their sides are all integers. We have already met the nicest of all (because its sides are small integers): the triangle with sides 3 units, 4 units and 5 units. But there are others.

Exercises

1. Show that a triangle with sides 6, 8, and 10 units is a right triangle.

2. Look at the exercises to section 3. These exercises use three more right triangles, all of whose sides are integers. Make a list of them. (Later, in Chapter 7, we will discover a way to find many more such right triangles.)

3. The legs of a right triangle are 8 and 15 units. Find the length of the hypotenuse.

4. We have used right triangles with the following sides:

Leg	Leg	Hypotenuse
3	4	5
6	8	10
9	12	15

   By continuing this pattern, find three more right triangles with integer sides.

5. We have seen that a triangle with sides 5, 12, and 13 is a right triangle. Can you find a right triangle, with integer sides, whose shortest side has length 10? length 15?

6. Exercises 4 and 5 suggest that we can construct one integer-side right triangle from another by multiplying each side by the same number (since the new triangle is *similar* to the old, it is still a right triangle). We can also reverse the process, dividing each side by the same number. Although we won't always get integers, we will always get *rational* numbers. Show that a triangle with sides 3/5, 4/5, and 1 is a right triangle.

7. Using the technique from Exercise 6, start with a 3-4-5 triangle and find a right triangle with rational sides whose shorter leg is 1. Then find a right triangle whose longer leg is 1.

8. Start with a 5-12-13 right triangle, and find a right triangle with rational sides whose hypotenuse is 1. Then find one whose shorter leg is 1. Finally, find a right triangle whose longer leg is 1.

9. Note that the right triangles with sides equal to 5, 12, 13 and 9, 12, 15 both have a leg equal to 12. Using this fact, find the area of a triangle with sides 13, 14, and 15.

10. (a) Find the area of a triangle with sides 25, 39, 56.
    (b) Find the area of a triangle with sides 25, 39, 16.

5 Our next best friends (among right triangles)

In the previous section, we explored right triangles with nice sides. We will now look at some triangles which have nice angles. For example, the two acute angles of the right triangle might be equal. Then the triangle is isosceles, and its acute angles are each 45°.

Or, we could take one acute angle to be double the other. Then the triangle has acute angles of 30 and 60°.

But nobody is perfect. It turns out that the triangles with nice angles never have nice sides. For example, in the case of the 45° right triangle, we have two equal legs, and a hypotenuse that is longer:

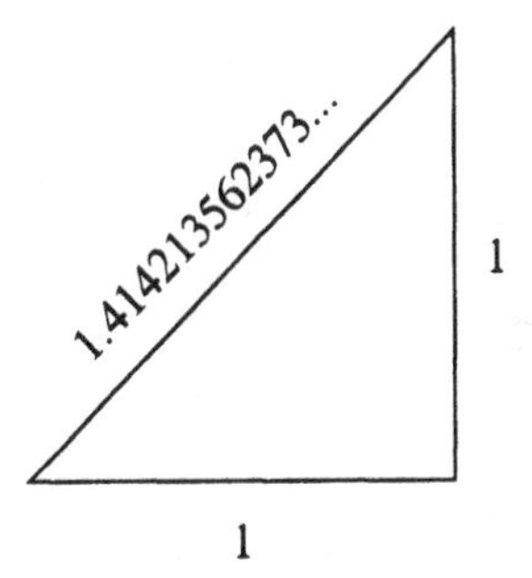

If we suppose the legs are each 1 unit long, then the hypotenuse, measured in the same units, is about 1.414213562373 units long, not a very nice number.

For a 30°right triangle, if the shorter leg is 1, the hypotenuse is a nice length[3]: it is 2. But the longer leg is not a nice length. It is approximately 1.732 (you can remember this number because its digits form the year in which George Washington was born – and the composer Joseph Haydn).

It also turns out that triangles with nice sides never have nice angles.

If we want an example of some theorem or definition, we will look at how the statement applies to our friendly triangles.

Exercises

1. Find the length of each leg of an isosceles right triangle whose hypotenuse has length 1. Challenge: Find the length, correct to nine decimal places without using your calculator (but using information contained in the text above!).

2. Using the Pythagorean theorem, find the hypotenuse of an isosceles right triangle whose legs are each three units long.

[3] If you don't remember the proof, just take two copies of such a triangle, and place them back-to-back:

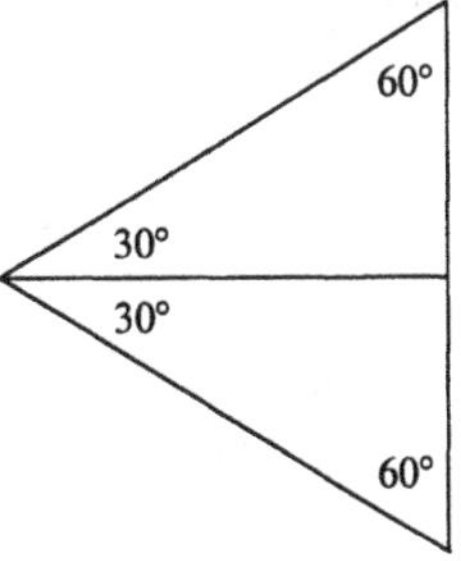

You will find that they form an equilateral triangle. The side opposite the 30°angle is half of one side of this equilateral triangle, and therefore half of the hypotenuse.

3. The shorter leg of a 30-60-90°triangle is 5 units. Using the Pythagorean theorem (and the facts about a 30°-60°-90° triangle referred to above), find the lengths of the other two sides of the triangle.

4. In each of the diagrams below, find the value of x and y:

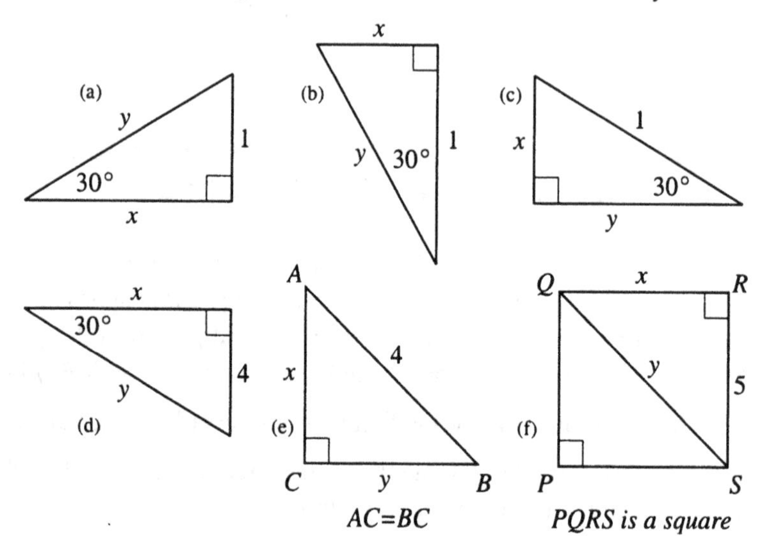

6 Some standard notation

A triangle has *six elements* ("parts"): three sides and three angles. We will agree to use capital letters, or small Greek letters, to denote the measures of the angles of the triangle (the same letters with which we denote the vertices of the angles). To denote the lengths of the sides of the triangle, we will use the small letter corresponding to the name of the angle opposite this side.

Some examples are given below:

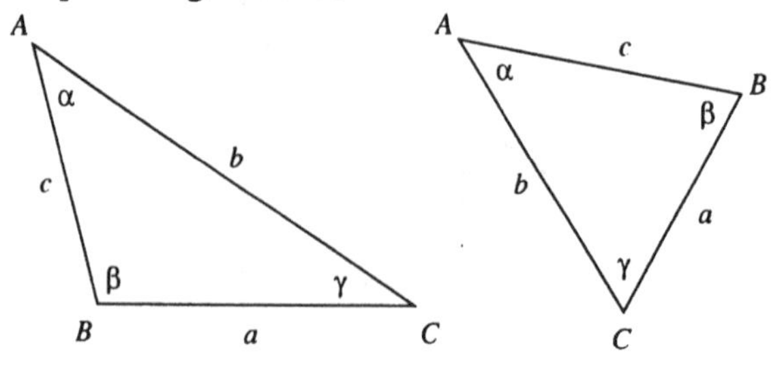

Appendix

I. Classifying triangles

Because the angles of any triangle add up to $180°$, a triangle can be classified as acute (having three acute angles), right (having one right angle), or obtuse (having one obtuse angle). We know from geometry that the lengths of the sides of a triangle determine its angles. How can we tell from these side lengths whether the triangle is acute, right, or obtuse?

Statement II of the Pythagorean theorem gives us a partial answer: If the side lengths a, b, c satisfy the relationship $a^2 + b^2 = c^2$, then the triangle is a right triangle. But what if this relationship is not satisfied?

We can tell a bit more if we think of a right triangle that is "hinged" at its right angle, and whose hypotenuse can stretch (as if made of rubber). The diagrams below show such a triangle. Sides a and b are of fixed length, and the angle between them is "hinged."

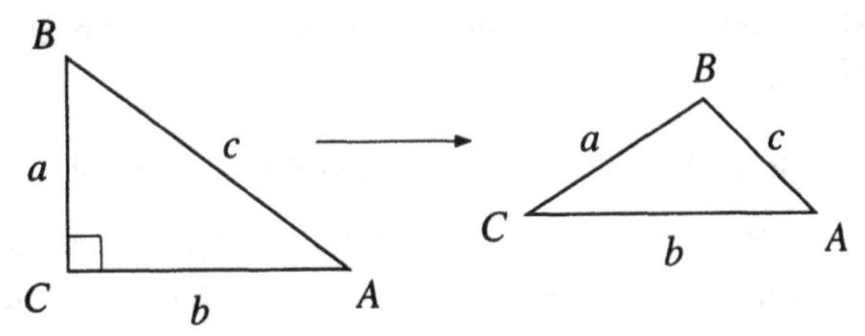

As you can see, if we start with a right triangle, and "close down" the hinge, then the right angle becomes acute. When this happens, the third side (labeled c) gets smaller. In the right triangle, $c^2 = a^2 + b^2$, so we can see that:

Statement III: If angle C of ΔABC is acute, then $c^2 < a^2 + b^2$.

In the same way, if we open the hinge up, angle C becomes obtuse, and the third side gets longer:

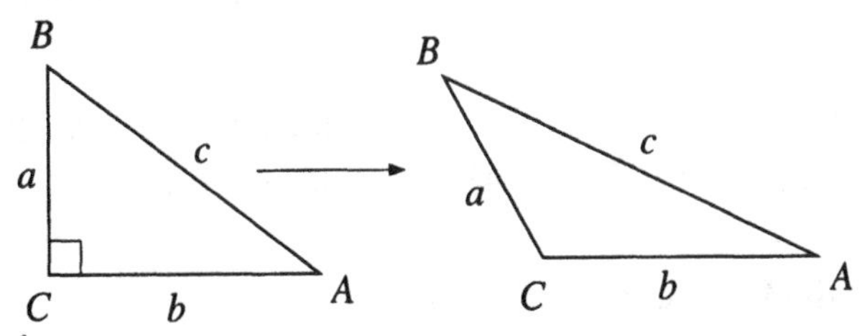

So we see that

Statement IV: If angle C of ΔABC is obtuse, then $c^2 > a^2 + b^2$.

Exercise Write the converse of statements III and IV above.

While the converses of most statements require a separate proof, for these particular cases, the converses follow from the original statements. For example, if, in $\triangle ABC$, $c^2 < a^2+b^2$, then angle C cannot be right (this would contradict statement II of the Pythagorean Theorem) and cannot be obtuse (this would contradict statement IV above). So angle C must be acute, which is what the converse of statement III says.

Statements III and IV, together with their converses, allow us to decide whether a triangle is acute, right, or obtuse, just by knowing the lengths of its sides.

Some examples follow:

1. Is a triangle with side lengths 2, 3, and 4 acute, right or obtuse?

   Solution. Since $4^2 = 16 > 2^2 + 3^2 = 4 + 9 = 13$, the triangle is obtuse, with the obtuse angle opposite the side of length 4.

   Question: Why didn't we need to compare 3^2 with $2^2 + 4^2$, or 2^2 with $3^2 + 4^2$?

2. Is a triangle with sides 4, 5, 6 acute, right, or obtuse?

   Solution. We need only check the relationship between 6^2 and $4^2 + 5^2$. Since $6^2 = 36 < 4^2 + 5^2 = 41$, the triangle is acute.

3. Is the triangle with side lengths 1, 2, and 3 acute, right, or obtuse?

   Solution. We see that $3^2 = 9 > 2^2 + 1^2 = 5$, so it looks like the triangle is obtuse.

   Question: This conclusion is *incorrect.* Why?

Exercise

If a triangle is constructed with the side lengths given below, tell whether it will be acute, right, or obtuse.

a) $\{6, 7, 8\}$ b) $\{6, 8, 10\}$ c) $\{6, 8, 9\}$ d) $\{6, 8, 11\}$
e) $\{5, 12, 12\}$ f) $\{5, 12, 14\}$ g) $\{5, 12, 17\}$

II. Proof of the Pythagorean theorem

There are many proofs of this classic theorem. Our proof follows the Greek tradition, in which the squares of lengths are interpreted as areas. We first recall statement I from the text:

If a and b are the lengths of the legs of a right triangle, and c is the length of its hypotenuse, then $a^2 + b^2 = c^2$.

Let us start with any right triangle. The lengths of its legs are a and b, and the length of its hypotenuse is c:

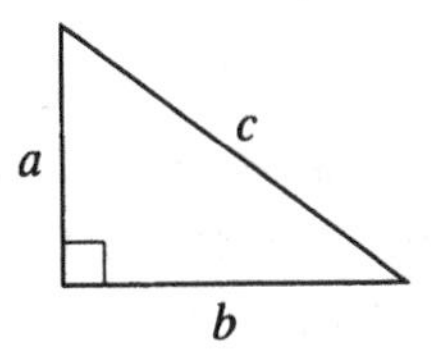

We draw a square (outside the triangle), on each side of the triangle:

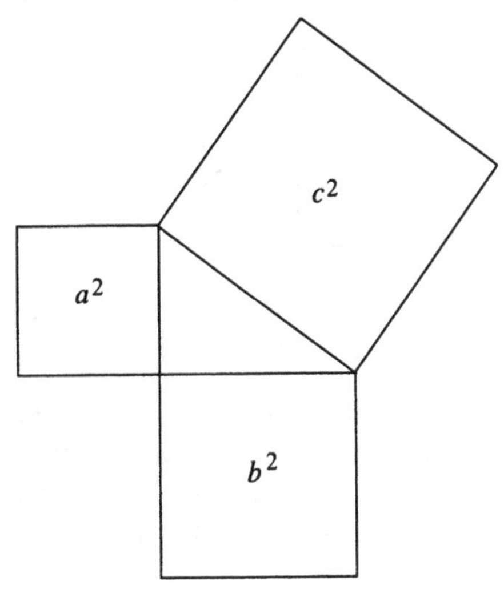

We must show that the sum of the areas of the smaller squares equals the area of the larger square:

$$a^2 + b^2 = c^2$$

Or:

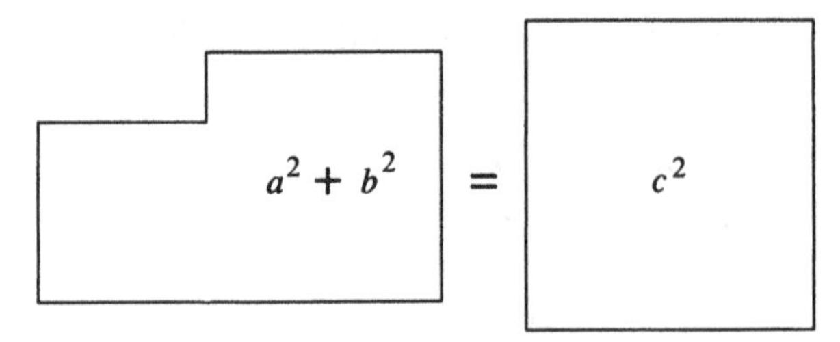

The diagram below gives the essence of the proof. If we cut off two copies of the original triangle from the first figure, and paste them in the correct niches, we get a square with side c:

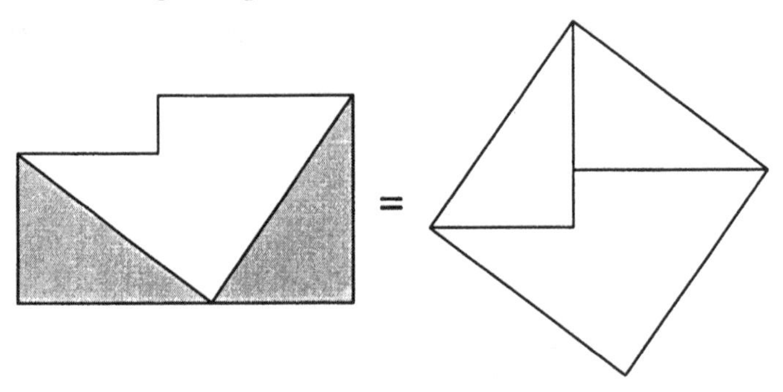

We fill in some details of the proof below.

We started with an oddly shaped hexagon, created by placing two squares together. To get the shaded triangle, we lay off a line segment equal to b, starting on the lower left-hand corner. Then we draw a diagonal line. This will leave us with a copy of the original triangle in the corner of the hexagon:

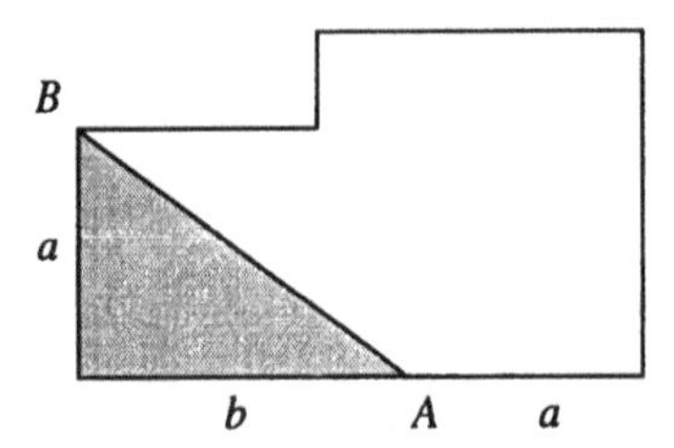

(Notice that the piece remaining along the bottom side of the hexagon has length a, since the whole bottom side had length $a + b$.)

Triangle ABC is congruent to the one we started with, because it has the same two legs, and the same right angle. Therefore hypotenuse AB will have length c.

Next we cut out the copy of our original triangle, and fit it into the niche created in our diagram:

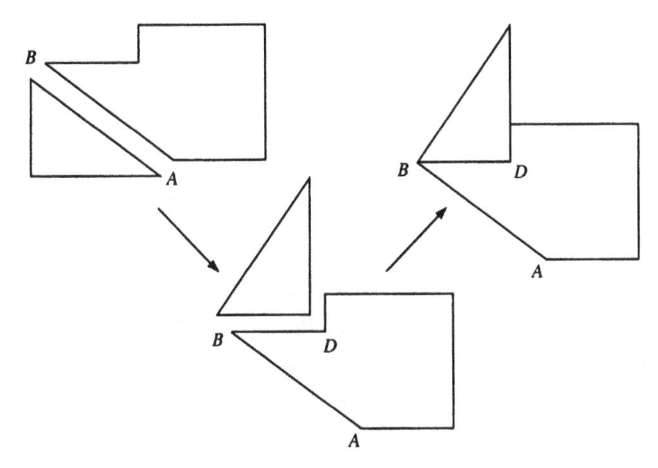

The right angle inside the triangle fits onto the right angle outside the hexagon (at D), and the leg of length a fits onto segment BD, which also has length a.

Connecting A to E, we form another triangle congruent to the original (we have already seen that $AF = a$, and $EF = b$ because each was a side of one of the original squares).

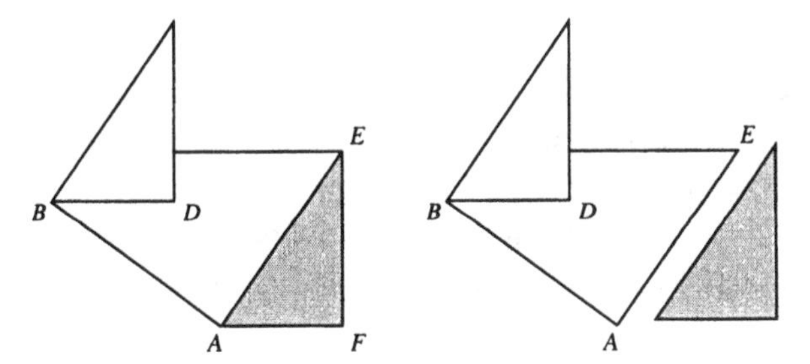

This new copy of the triangle will fit nicely in the niche created at the top of the diagram:

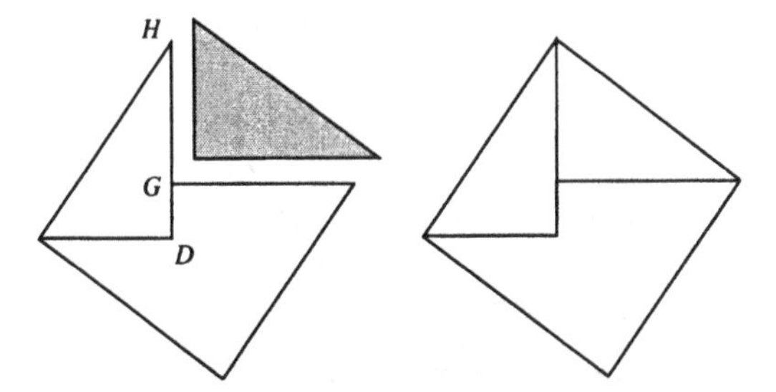

Why will it fit? The longer leg, of length b, is certainly equal to the upper side of the original hexagon. And the right angles at G must fit together. But why does GH fit with the other leg of the triangle, which is of length a?

Let us look again at the first copy of our original triangle. If we had placed it alongside the square of side b, it would have looked like this:

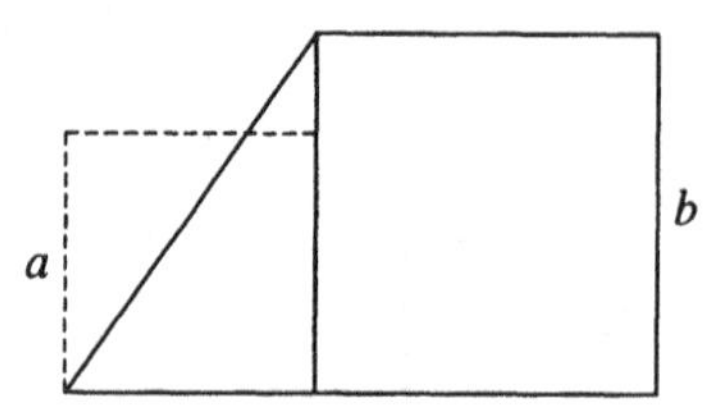

But in fact we draw it sitting on top of the smaller square, so it was pushed up vertically by an amount equal to the side of this square, which is a:

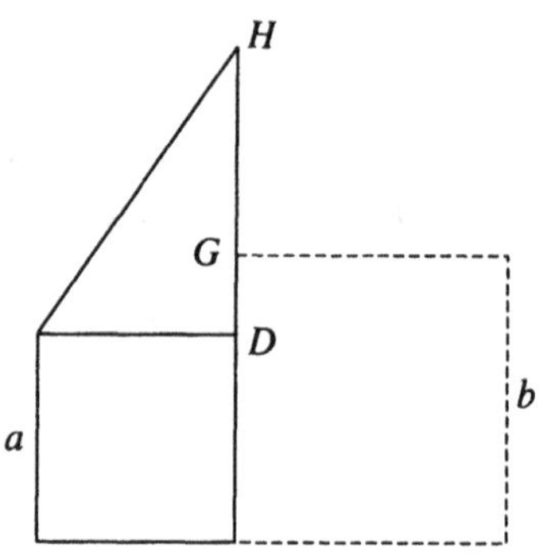

So the amount that it protrudes above point G must be equal to a. This is the length of GH, which must then fit with the smaller leg of the second copy of our triangle.

One final piece remains: why is the final figure a square? Certainly, it has four sides, all equal to length c. But why are its angles all right angles?

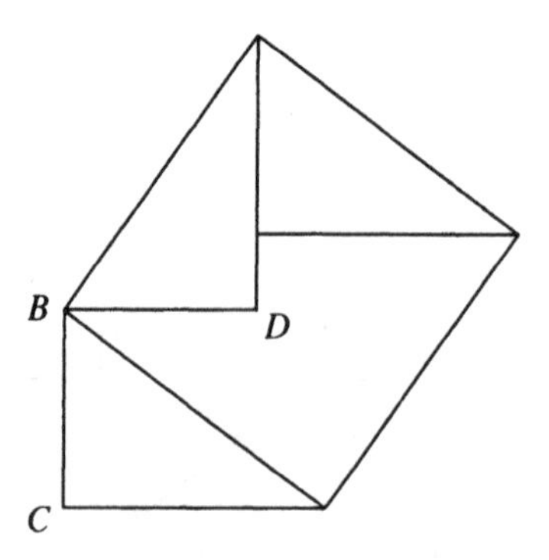

Let us look, for example, at vertex B. Angle CBD was originally a right angle (it was an angle of the smaller square). We took a piece of it away when we cut off our triangle, and put the same piece back when we pasted the triangle back in a different position. So the new angle, which is one in our new figure, is still a right angle. Similar arguments hold for other vertices in our figure, so it must be a square.

In fact all the pieces of our puzzle fit together, and we have transformed the figure consisting of squares with sides a and b into a square with side c. Since we have not changed the area of the figure, it must be true that a $a^2 + b^2 = c^2$.

Finally, we prove statement II of the text:

> If the positive numbers a, b, and c satisfy $a^2 + b^2 = c^2$, then a triangle with these side lengths has a right angle opposite the side with length c.

We prove this statement in two parts. First we show that the numbers a, b, and c are sides of some triangle, then we show that the triangle we've created is a right triangle.

Geometry tells us that three numbers can be the sides of a triangle if and only if the sum of the smallest two of them is greater than the largest. But can we tell which of our numbers is the largest? We can, if we remember that for *positive* numbers, $p^2 > q^2$ implies that $p > q$. Since $c^2 = a^2 + b^2$, and $b^2 > 0$, we see that $c^2 > a^2$, so $c > a$. In the same way, we see that $c > b$.

Now we must show that $a + b > c$. Again, we examine the squares of our numbers. We find that $(a+b)^2 = a^2 + 2ab + b^2 > a^2 + b^2 = c^2$ (since $2ab$ is a positive number). So $a + b > c$ and segments of lengths a, b, and c form a triangle.

What kind of triangle is it? Let us draw a picture:

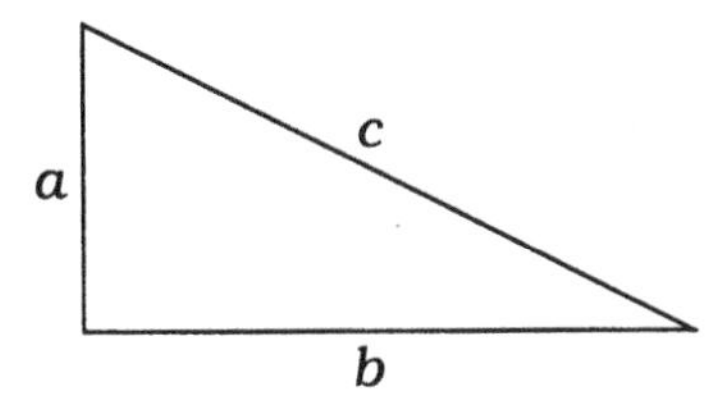

Does this triangle contain a right angle? We can test to see if it does by copying parts of it into a new triangle. Let us draw a new triangle with sides a and b, and a right angle between them:

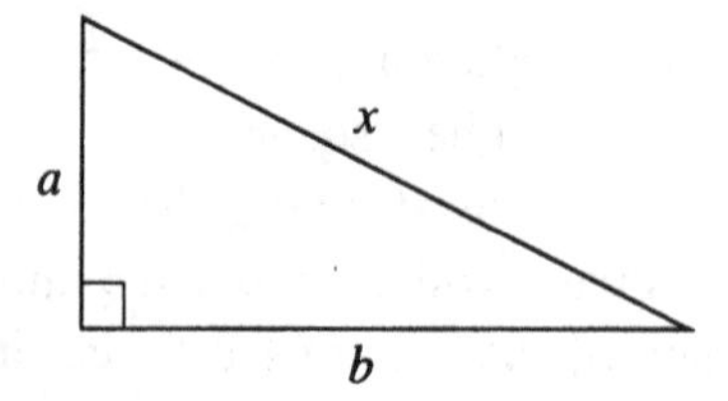

How long is the hypotenuse of this new triangle? If its length is x, then statement I of the Pythagorean theorem (which we have already proved) tells us that $x^2 = a^2 + b^2$. But this means that $x^2 = c^2$, or $x = c$. It remains to note that this new triangle, which has the same three sides as the original one, is congruent to it. Therefore the sides of length a and b in our original triangle must contain a right angle, which is what we wanted to prove.

Chapter 1

Trigonometric Ratios in a Triangle

1 Definition of $\sin\alpha$

Definition: For any acute angle α, we draw a right triangle that includes α. The *sine* of α, abbreviated $\sin\alpha$, is the ratio of the length of the leg opposite this angle to the length of the hypotenuse of the triangle.

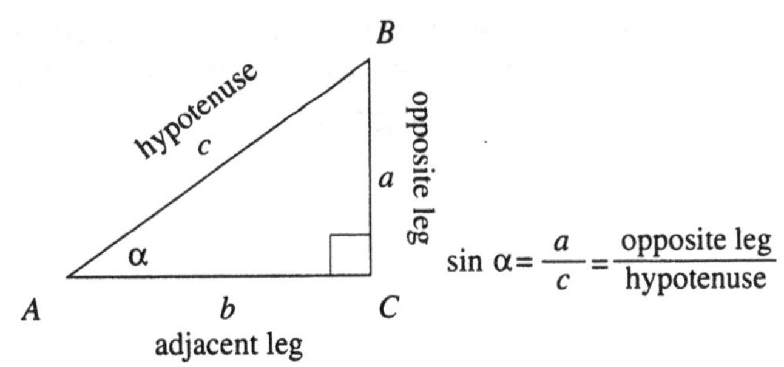

For example, in the right triangle ABC (diagram above), $\sin\alpha = a/c$. We can see immediately that this definition has a weak point: it does not tell us exactly which right triangle to draw. There are many right triangles, large ones and small ones that include a given angle α.

Let us try to answer the following questions.

Example 12 Find $\sin 30°$.

"Solution" 1. Formally, we are not obliged to solve the problem, since we are given only the measure of the angle, without a right triangle that includes it. □

Solution 2. Draw some right triangle with a 30° angle:

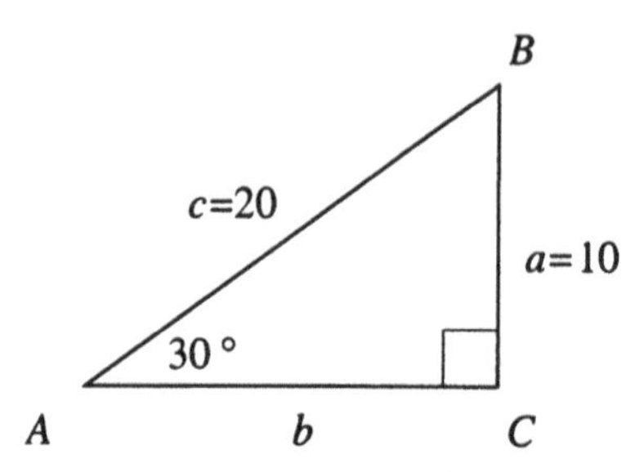

For example, we might let the length of the hypotenuse be 20. Then the length of the side opposite the 30° angle measures 10 units. So

$$\sin 30^\circ = \frac{10}{20} = \frac{1}{2} = 0.5\,.$$

We know, from geometry, that whatever the value of the hypotenuse, the side opposite the 30° angle will be half this value,[1] so $\sin 30^\circ$ will always be $1/2$. This value depends only on the measure of the angle, and not on the lengths of the sides of the particular triangle we used. □

Example 13 An American student is writing by e-mail to her friend in France, and they are doing homework together. The American student writes to the French student: "Look at page 22 of the Gelfand–Saul Trigonometry book. Let's get the sine of angle D."

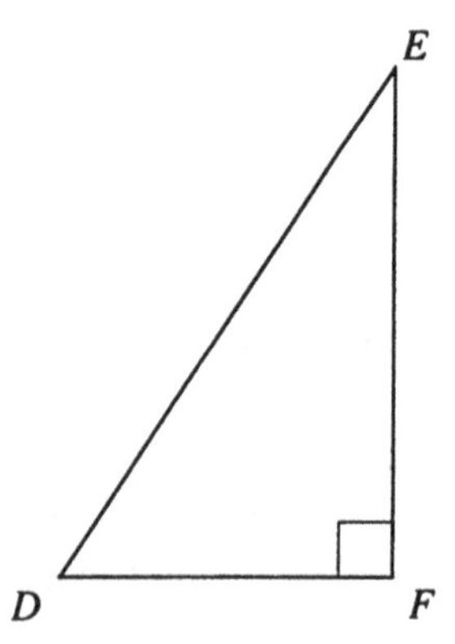

The French student measured EF with his ruler, then measured ED, then took the ratio EF/ED and sent the answer to his American friend. A

[1] A theorem in geometry tells us that in a right triangle with a 30° angle the side opposite this angle is half the hypotenuse (see Chapter 0, page 11).

few days later, he woke up in the middle of the night and realized, "*Sacré bleu!* I forgot that Americans use inches to measure lengths, while we use centimeters. I will have to tell my friend that I gave her the wrong answer!" What must the French student do to correct his answer?

Solution. He does not have to do anything – the answer is correct. The sine of an angle is a ratio of two lengths, which does not depend on any unit of measurement. For example, if one segment is double another when measured in centimeters, it is also double the other when measured in inches. □

In general, for any angle of $\alpha°$ (for $0 < \alpha < 90$), the value of $\sin\alpha$ depends only on α, and not on the right triangle containing the angle.[2] This is true because any two triangles containing acute angle α are similar, so the ratios of corresponding sides are equal. Sin α is merely a name for one of these ratios.

Exercises

1. In each diagram below, what is the value of $\sin\alpha$?

a)

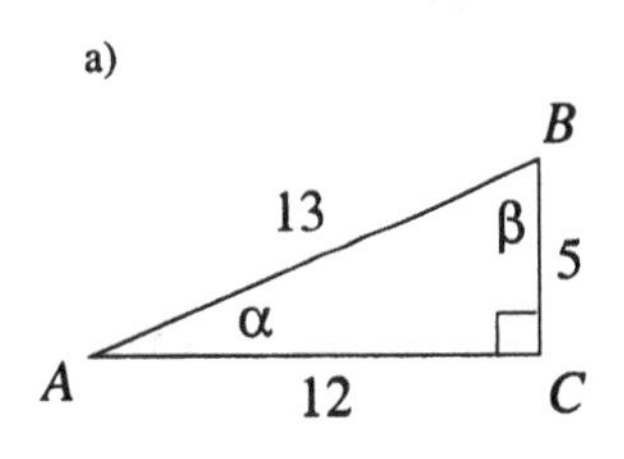

b)

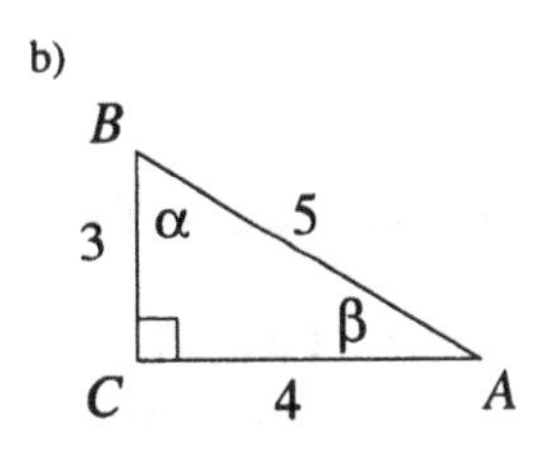

c)

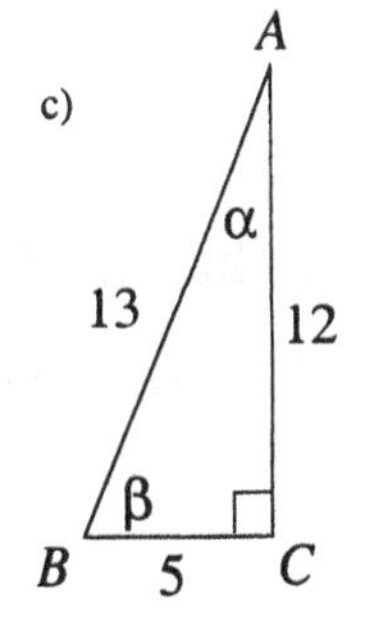

[2]Example 12 shows that the value of $\sin\alpha$ does not depend on the particular triangle which contains α. Example 13 shows that the value of $\sin\alpha$ does not depend on the unit of measurement for the sides of the triangle. In fact, we can examine Example 12 more closely. To determine the value of $\sin 30°$, we need three pieces of information: (a) the angle; (b) the right triangle containing the angle; (c) the unit of measurement for the sides of the triangle. We have just shown that the value of $\sin\alpha$ does not in fact depend on the last two pieces of information.

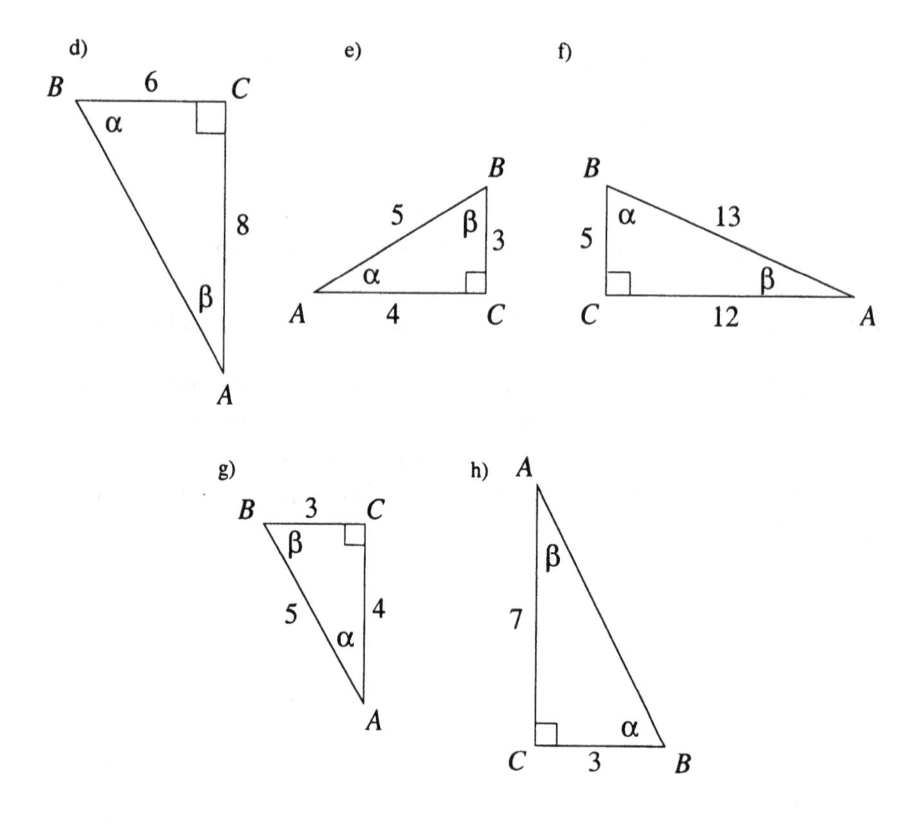

2. In each of the diagrams above, find $\sin \beta$.

3. In the following list, cross off each number which is less than the sine of 60°. Then check your work with a calculator.

   0.1 0.2 0.3 0.4 0.5 0.6 0.7 0.8 0.9

   Hint: Remember the relationships among the sides of a 30-60-90 triangle.

2 Find the hidden sine

Sometimes the sine of an angle lurks in a diagram where it is not easy to spot. The following exercises provide practice in finding ratios equal to the sine of an angle, and lead to some interesting formulas.

Exercises

1. The diagram below shows a right triangle with an altitude drawn to the hypotenuse. The small letters stand for the lengths of certain line segments.

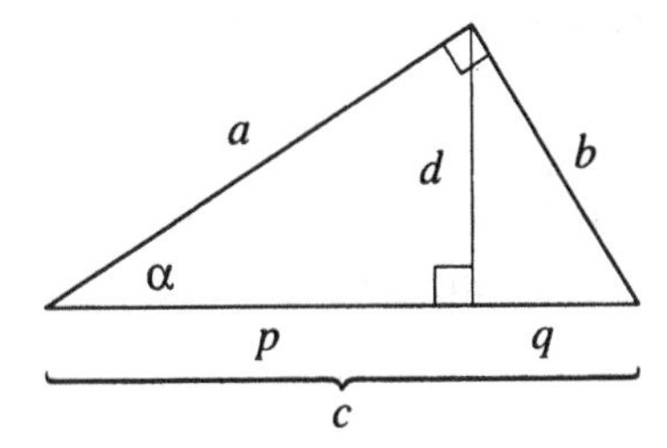

   a) Find a ratio of the lengths of two segments equal to $\sin\alpha$.
   b) Find another ratio of the lengths of two segments equal to $\sin\alpha$.
   c) Find a third ratio of the lengths of two segments equal to $\sin\alpha$.

2. The three angles of triangle ABC below are acute (in particular, none of them is a right angle), and CD is the altitude to side AB. We let $CD = h$, and $CA = b$.

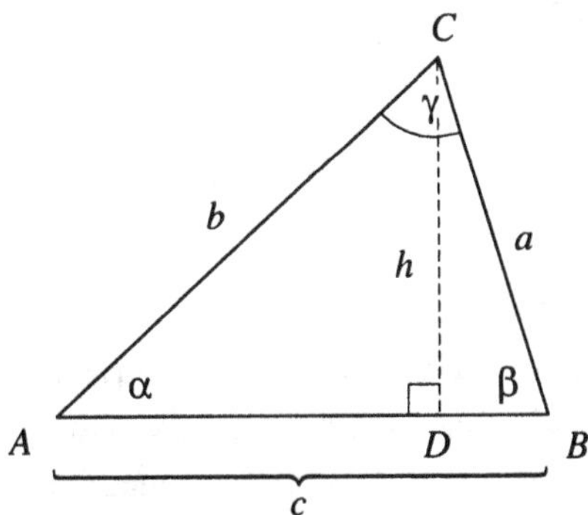

   a) Find a ratio equal to $\sin\alpha$.
   b) Express h in terms of $\sin\alpha$ and b.
   c) We know that the area of triangle ABC is $hc/2$. Express this area in terms of b, c, and $\sin\alpha$.
   d) Express the area of triangle ABC in terms of a, c, and $\sin\beta$.
   e) Express the length of the altitude from A to BC in terms of c and $\sin\beta$. (You may want to draw a new diagram, showing the altitude to side BC.)

3. a) Using the diagram above, write two expressions for h: one using side b and $\sin\alpha$ and one using side a and $\sin\beta$.
   b) Using the result to part (a), show that $a\sin\beta = b\sin\alpha$.
   c) Using the result of part (e) in problem 2 above, show that $c\sin\beta = b\sin\gamma$.
   d) Prove that $\frac{a}{\sin\alpha} = \frac{b}{\sin\beta} = \frac{c}{\sin\gamma}$. This relation is true for any acute triangle (and, as we will see, even for any obtuse triangle). It is called the *Law of Sines*.

3 The cosine ratio

Definition: In a right triangle with acute angle α, the ratio of the leg adjacent to angle α to the hypotenuse is called the *cosine* of angle α, abbreviated $\cos\alpha$.

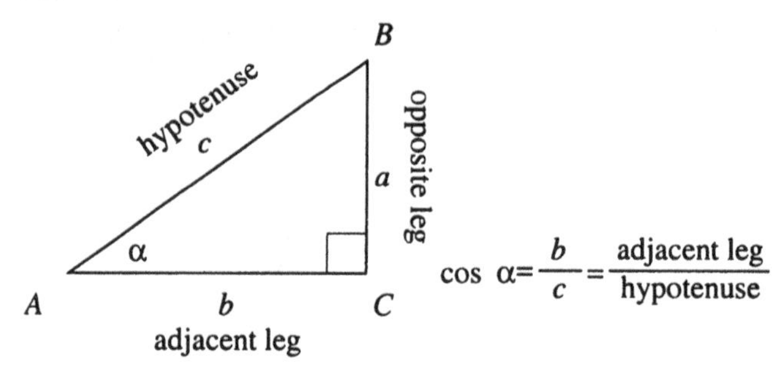

Notice that the value of $\cos\alpha$, like that of $\sin\alpha$, depends only on α and not on the right triangle that includes α. Any two such triangles will be similar, and the ratio $\cos\alpha$ will thus be the same in each.

Exercises

1. Find the cosines of angles α and β in each of the triangle figures in Exercise 1 beginning on page 23.

2. Find the cosines of angles α and β in each triangle below.

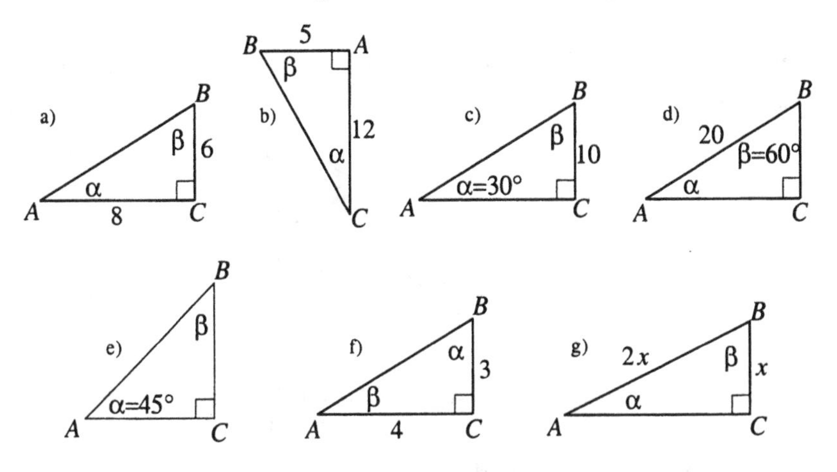

3. The diagram below shows a right triangle with an altitude drawn to the hypotenuse. The small letters stand for the lengths of certain line segments.

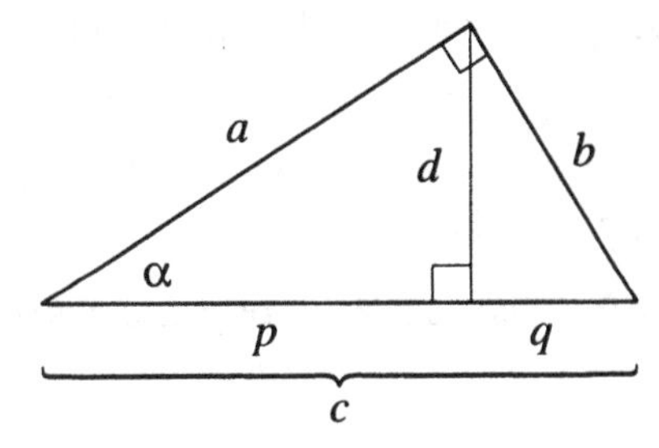

a) Find a ratio of the lengths of two segments equal to $\cos\alpha$.

b) Find another ratio of the lengths of two segments equal to $\cos\alpha$.

c) Find a third ratio of the lengths of two segments equal to $\cos\alpha$.

4 A relation between the sine and the cosine

Example 14 In the following diagram, $\cos\alpha = 5/7$. What is the numerical value of $\sin\beta$?

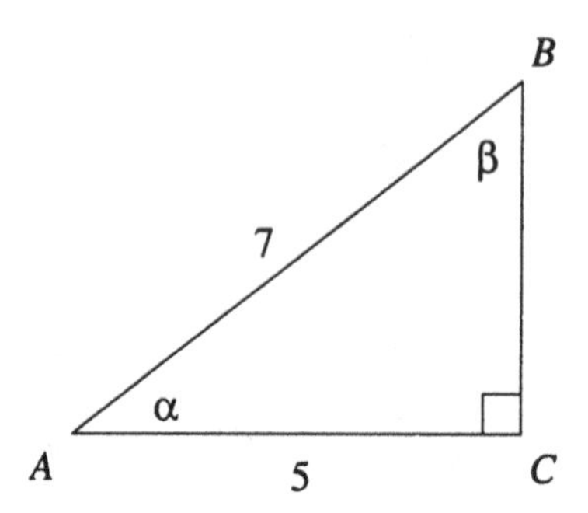

Solution. By the definition of the sine ratio

$$\sin\beta = \frac{AC}{AB}.$$

The value of this ratio is $5/7$, which is the same as $\cos\alpha$. □

Is this a coincidence? Certainly not. If α and β are acute angles of the same right triangle, $\sin\alpha = \cos\beta$, no matter what lengths the sides of the triangle may have. We state this as a

Theorem If $\alpha + \beta = 90°$, then $\sin\alpha = \cos\beta$ and $\cos\alpha = \sin\beta$.

Exercises

1. Show that $\sin 29° = \cos 61°$.

2. If $\sin 35° = \cos x$, what could the numerical value of x be?

3. Show that we can rewrite the theorem of the above section as: $\sin\alpha = \cos(90 - \alpha)$.

5 A bit of notation

If we are not careful, ambiguity arises in certain notation. What does $\sin x^2$ mean? Do we square the angle, then take its sine? Or do we take $\sin x$ first, then square this number? The first case is very rare: why should we want to square an angle? What units could we use to measure such a quantity?

The second case happens very often. Let us agree to write $\sin^2 x$ for $(\sin x)^2$, which is the case where we take the sine of an angle, then square the result. For example, $\sin^2 30° = 1/4$.

Exercise

In the diagram below, find the numerical value of the following expressions:

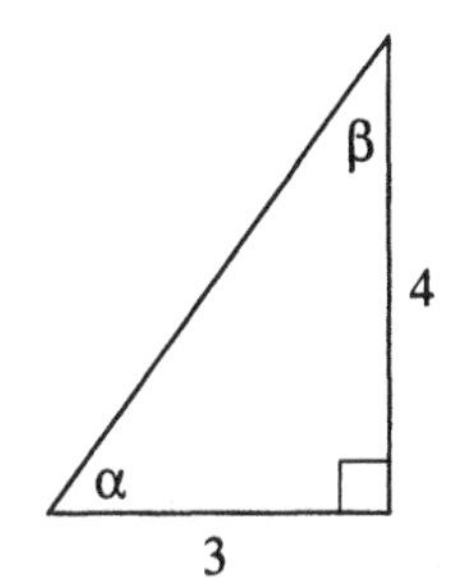

1. $\sin^2 \alpha$
2. $\sin^2 \beta$
3. $\cos^2 \alpha$
4. $\cos^2 \beta$
5. $\sin^2 \alpha + \cos^2 \alpha$
6. $\sin^2 \alpha + \cos^2 \beta$
7. $\cos^2 \alpha + \sin^2 \beta$

6 Another relation between the sine and the cosine

If you look carefully among the exercises of the previous section, you will see examples of the following result:

Theorem For any acute angle α, $\sin^2 \alpha + \cos^2 \alpha = 1$.

Proof As usual, we draw a right triangle that includes the angle α:

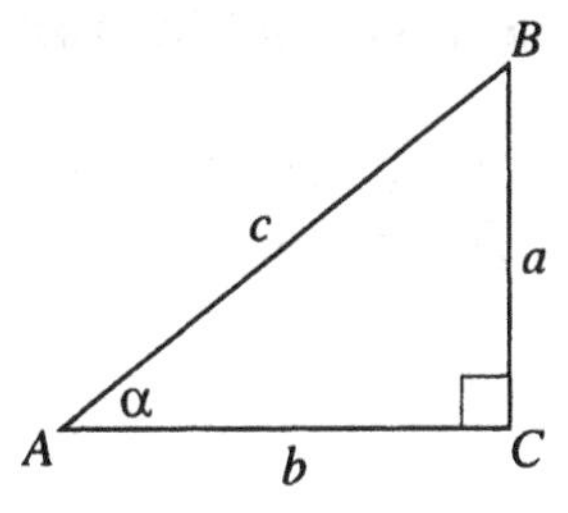

Suppose the legs have lengths a and b, and the hypotenuse has length c. Then $\sin^2 \alpha + \cos^2 \alpha = (a/c)^2 + (b/c)^2 = (a^2 + b^2)/c^2$. But the Pythagorean theorem tells us that $a^2 + b^2 = c^2$, so the last fraction is equal to 1; that is, $\sin^2 \alpha + \cos^2 \alpha = 1$. □

Exercises

1. Verify that $\sin^2\alpha + \cos^2\alpha = 1$, where α is the angle in the following diagram:

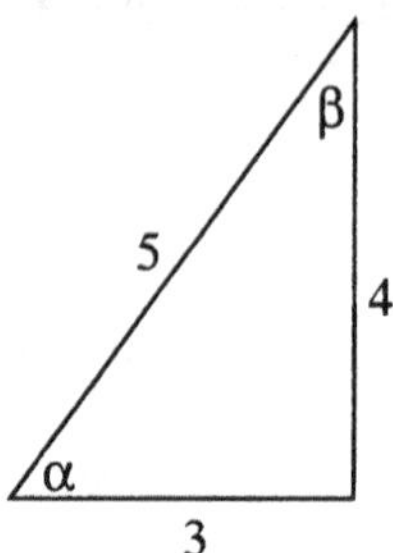

2. Did you notice that no right angle is indicated in the diagram above? Is that an error?

3. Verify that $\sin^2\beta + \cos^2\beta = 1$, where β is the other angle in the diagram.

4. Find the value of $\cos\alpha$ if α is an acute angle and $\sin\alpha = 5/13$.

5. Find the value of $\cos\alpha$ if α is an acute angle and $\sin\alpha = 5/7$.

6. If α and β are acute angles in the same right triangle, show that $\sin^2\alpha + \sin^2\beta = 1$.

7. If α and β are acute angles in the same right triangle, show that $\cos^2\alpha + \cos^2\beta = 1$.

7 Our next best friends (and the sine ratio)

It is usually not very easy to find the sine of an angle, given its measure. But for some special angles, it is not so difficult. We have already seen that $\sin 30^\circ = 1/2$.

Example 15 Find $\cos 30^\circ$.

Solution. To use our definition of the cosine of an angle, we must draw a right triangle with a 30° angle, a triangle with which we are already friendly.

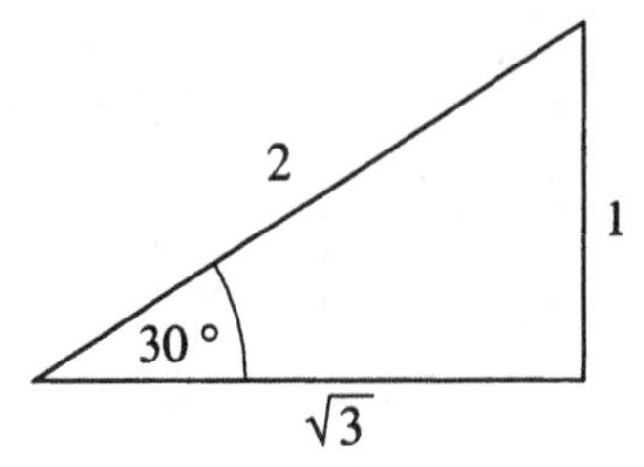

We know that the value of $\cos 30^\circ$ depends only on the shape of this triangle, and not on its sides. So we can assume that the smaller leg has length 1. Then the lengths of the other sides are as shown in the diagram, and we see that $\cos 30^\circ = \sqrt{3}/2$. □

Example 16 Show that $\cos 60^\circ = \sin 30^\circ$.

Solution. In the 30-60-90 triangle we've drawn above, one acute angle is 30°, and the other is 60°. Standing on the vertex of the 30° angle, we see that the opposite leg has length 1, and the hypotenuse has length 2. Thus $\sin 30^\circ = 1/2$. But if we walk over to the vertex of the 60° angle, the opposite leg becomes the adjacent leg, and we see that the ratio that was $\sin 30^\circ$ earlier is also $\cos 60^\circ$. □

Exercises

1. Fill in the following table. You may want to use the model triangles given in the diagram below.

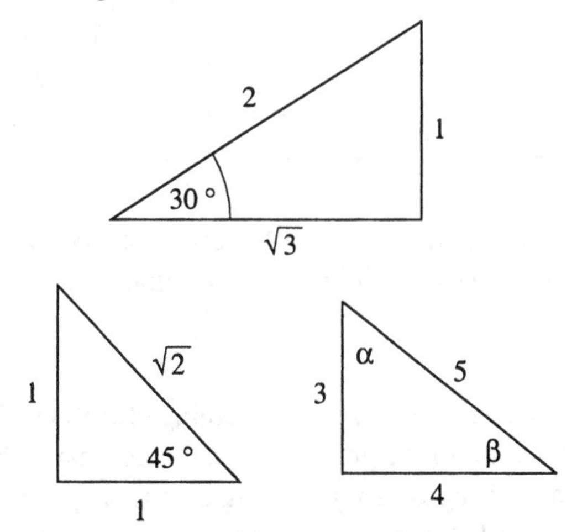

(The angles α and β are angles in a 3-4-5 right triangle.)

angle x	$\sin x$	$\cos x$
30°		
45°		
60°		
α		
β		

2. Verify that $\sin 60^\circ = \cos 30^\circ$.

3. Verify that $\sin^2 30^\circ + \cos^2 30^\circ = 1$.

4. Let the measure of the smaller acute angle in a 3-4-5 triangle be α. Looking at the values for $\sin\alpha$ and $\cos\alpha$, how large would you guess α is? Is it larger or smaller than 30°? Than 45°? Than 60°?

8 What is the value of sin 90°?

So far we have no answer to this question: We defined $\sin\alpha$ only for an acute angle. But there is a reasonable way to define $\sin 90^\circ$. The picture below shows a series of triangles with the same hypotenuse, but with different acute angles α:

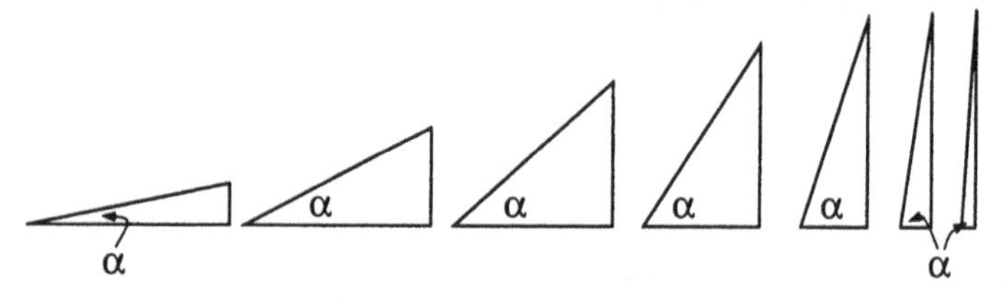

As the angle α gets larger, the ratio of the opposite side to the hypotenuse approaches 1. So we make the following definition.

Definition $\sin 90^\circ = 1$.

The diagram above also suggests something else about $\sin\alpha$. Remember that the hypotenuse of a right triangle is longer than either leg. Since $\sin\alpha$ is the ratio of a leg of a right triangle to its hypotenuse, $\sin\alpha$ can never be larger than 1. So if someone tells you that, for a certain angle α, $\sin\alpha = 1.2$ or even 1.01, you can immediately tell him or her that a mistake has been made.

The same series of triangles lets us make a definition for $\cos 90°$. As the angle α gets closer and closer to $90°$, the hypotenuse remains the same length, but the adjacent leg gets shorter and shorter. This same diagram leads us to the following definition.

Definition $\cos 90° = 0$.

Exercises

1. How does the diagram above lead us to make the definition that $\sin 0° = 0$?
2. What definition does the diagram in this section suggest for $\cos 0°$?

   Answer: $\cos 0° = 1$.
3. Check, using our new definitions, that $\sin^2 0° + \cos^2 0° = 1$.
4. Check, using our new definitions, that $\sin^2 90° + \cos^2 90° = 1$.
5. Your friend tells you that he has calculated the cosine of a certain angle, and his answer is 1.02. What should you tell your friend?

9 An exploration: How large can the sum be?

We have seen that the value of the expression $\sin^2 \alpha + \cos^2 \alpha$ is always 1. Let us now look at the expression $\sin \alpha + \cos \alpha$. What values can this expression take on? This question is not a simple one, but we can start thinking about it now.

Exercises

1. We can ask our best friends for some information. Fill in the blank spaces in the following table, using a calculator when necessary .

$\sin 0° + \cos 0°$	$0 + 1$	1
$\sin 30° + \cos 30°$		
$\sin 45° + \cos 45°$		
$\sin 60° + \cos 60°$	$\frac{\sqrt{3}}{2} + \frac{1}{2}$	1.366 (approximately)
$\sin 90° + \cos 90°$		
$\sin\alpha + \cos\alpha$, where α is the smaller acute angle in a 3-4-5 right triangle	$\frac{3}{5} + \frac{4}{5}$	1.4
$\sin\alpha + \cos\alpha$, where α is the larger acute angle in a 3-4-5 right triangle		

2. Prove that $\sin\alpha + \cos\alpha$ is always less than 2.

   Hint: Geometry tells us that $\sin\alpha \leq 1$ and $\cos\alpha \leq 1$. Can they both be equal to 1 for the same angle?

3. Show that $\sin\alpha + \cos\alpha \geq 1$ for any acute angle α.

   Hint: Notice that $(\sin\alpha + \cos\alpha)^2 = 1 + 2\sin\alpha\cos\alpha$, and think about how this shows what we wanted.

4. For what value of α is $\sin\alpha + \cos\alpha = \sqrt{2}$?

5. We can see, from the table above, that $\sin\alpha + \cos\alpha$ can take on the value 1.4. Can it take the value 1.5? We will return to this problem a bit later. For now, use your calculator to see how large a value you can get for the expression $\sin\alpha + \cos\alpha$.

10 More exploration: How large can the product be?

Now let us consider the product $(\sin\alpha)(\cos\alpha)$. How large can this be?

Fill in the table below:

$(\sin 0°)(\cos 0°)$	$0 \cdot 1$	0
$(\sin 30°)(\cos 30°)$		
$(\sin 45°)(\cos 45°)$		
$(\sin 60°)(\cos 60°)$		0.43301 (approximately)
$(\sin 90°)(\cos 90°)$		
$(\sin\alpha)(\cos\alpha)$, where α is the smaller acute angle in a 3-4-5 right triangle	$\frac{3}{5} \cdot \frac{4}{5}$	0.48
$(\sin\alpha)(\cos\alpha)$, where α is the larger acute angle in a 3-4-5 right triangle		

How large do you think the product $(\sin\alpha)(\cos\alpha)$ can get? We will return to this problem later on.

11 More names for ratios

In a right triangle,

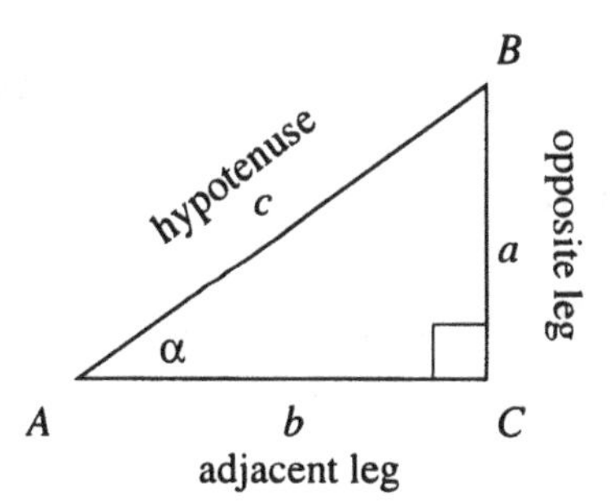

we have a total of six different ratios of sides. Each of these ratios has been given a special name.

We have already given a name to two of these ratios:

$$\sin\alpha = \frac{a}{c}, \qquad \cos\alpha = \frac{b}{c}.$$

Below we give names for the remaining four ratios. The first two are very important.

The ratio

$$\frac{\text{leg opposite angle } \alpha}{\text{leg adjacent to angle } \alpha} = \frac{a}{b} \qquad \boxed{\tan\alpha = \frac{\text{opposite leg}}{\text{adjacent leg}}}$$

is called the *tangent* of α, abbreviated $\tan\alpha$.

The ratio

$$\frac{\text{leg adjacent to angle } \alpha}{\text{leg opposite angle } \alpha} = \frac{b}{a} \qquad \boxed{\cot\alpha = \frac{\text{adjacent leg}}{\text{opposite leg}}}$$

is called the *cotangent* of α, abbreviated $\cot\alpha$.

Two more ratios are used in some textbooks, but are not so important mathematically. We list them here for completeness, but will not be working with them:

The ratio

$$\frac{\text{hypotenuse}}{\text{leg adjacent to angle } \alpha} = \frac{c}{b} \qquad \boxed{\sec\alpha = \frac{\text{hypotenuse}}{\text{adjacent leg}}}$$

is called the *secant* of α, abbreviated $\sec\alpha$.

The ratio

$$\frac{\text{hypotenuse}}{\text{leg opposite angle } \alpha} = \frac{c}{a} \qquad \boxed{\csc\alpha = \frac{\text{hypotenuse}}{\text{opposite leg}}}$$

is called the *cosecant* of α, abbreviated $\csc\alpha$.

For practice, let's take the example of a 3-4-5 right triangle:

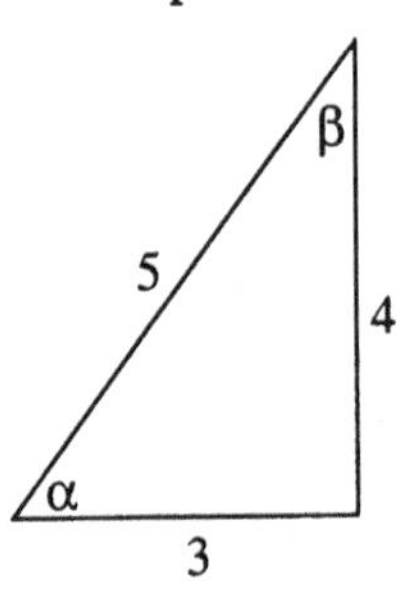

We have six ratios and six names:

$$\sin\alpha = \frac{4}{5} \qquad \cos\alpha = \frac{3}{5} \qquad \tan\alpha = \frac{4}{3}$$

$$\cot\alpha = \frac{3}{4} \qquad \sec\alpha = \frac{5}{3} \qquad \csc\alpha = \frac{5}{4}$$

To sum up, given the triangle

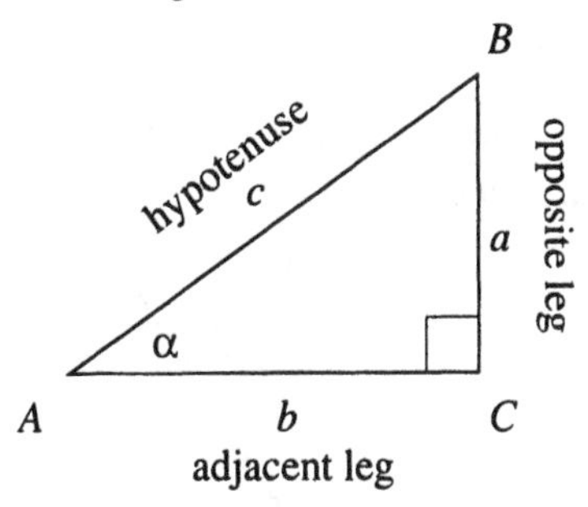

we have:

$\sin\alpha$	opposite leg/hypotenuse	a/c
$\cos\alpha$	adjacent leg/hypotenuse	b/c
$\tan\alpha$	opposite leg/adjacent leg	a/b
$\cot\alpha$	adjacent leg/opposite leg	b/a
$\sec\alpha$	hypotenuse/adjacent leg	c/b
$\csc\alpha$	hypotenuse/opposite leg	c/a

As before, these ratios depend only on the size of the angle α, and not on the lengths of the sides of the particular triangle we are using, or on how we measure the sides. The following theorem generalizes our statement of this fact for $\sin\alpha$.

Theorem The values of the trigonometric ratios of an acute angle depend only on the size of the angle itself, and not on the particular right triangle containing the angle.

Proof Any two triangles containing a given acute angle are similar, so ratios of corresponding sides are equal. The trigonometric ratios are just names for these ratios. □

Exercises

1. For the angles in the figure below, find $\cos\alpha$, $\cos\beta$, $\sin\alpha$, $\sin\beta$, $\tan\alpha$, $\tan\beta$, $\cot\alpha$ and $\cot\beta$.

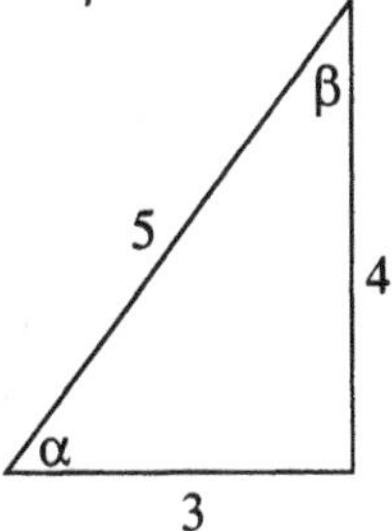

2. Did you assume that the triangle in the figure above is a right triangle? Why is this assumption correct?

3. Given the figure below, express the quantities $\cos\alpha$, $\cos\beta$, $\sin\alpha$, $\sin\beta$, $\tan\alpha$, $\tan\beta$, $\cot\alpha$ and $\cot\beta$ in terms of a, b, and c:

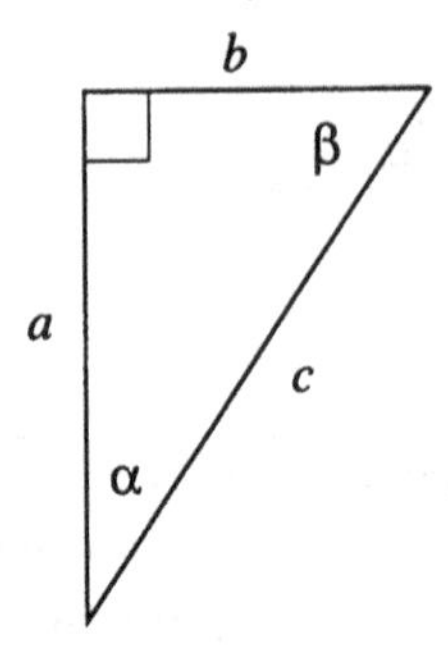

4. In the diagram below, find the numerical value of $\cos\alpha$, $\cos\beta$, $\cot\alpha$ and $\cot\beta$.

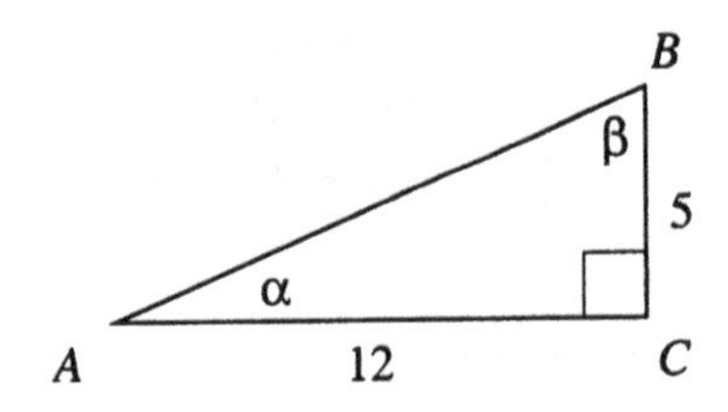

5. In the diagram below, find the numerical value of $\cos\alpha$, $\cos\beta$, $\cot\alpha$ and $\cot\beta$.

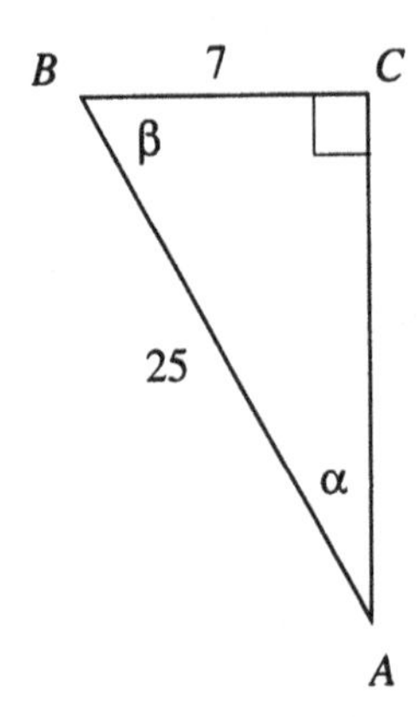

6. Find two names for each ratio given. The first example is already filled in.

$$\frac{a}{c} = \underline{\sin\alpha} = \underline{\cos\beta}$$

$$\frac{b}{c} = \underline{\quad\quad} = \underline{\quad\quad}$$

$$\frac{a}{b} = \underline{\quad\quad} = \underline{\quad\quad}$$

$$\frac{b}{a} = \underline{\quad\quad} = \underline{\quad\quad}$$

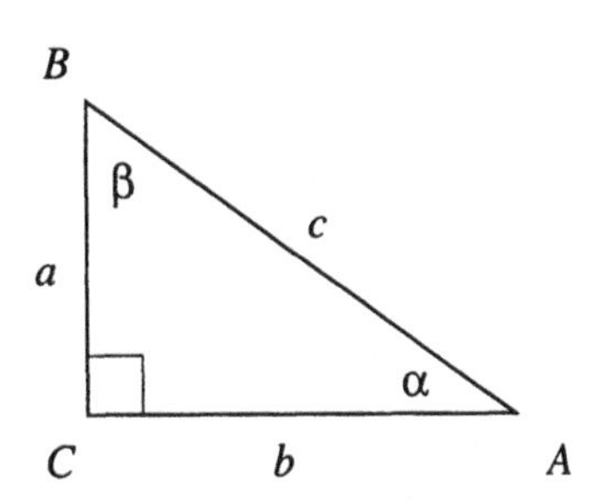

7. The sine of an angle is 3/5. What is its cosine? What is its cotangent?

8. If $\tan\alpha = 1$, what is $\cos\alpha$? What is $\cot\alpha$?

9. What is the numerical value of $\tan 45°$?

10. What is the numerical value of $\tan 30°$? Express this number using radicals. Then use a calculator to get an approximate numerical value.

11. What is the numerical value of $\tan 45° + \sin 30°$? Why don't you need a calculator to compute this?

Chapter 2

Relations among Trigonometric Ratios

1 The sine and its relatives

We have studied four different trigonometric ratios: sine, cosine, tangent, and cotangent. These four are closely related, and it will be helpful to explore their relationships. We have already seen that $\sin^2\alpha + \cos^2\alpha = 1$, for any acute angle α. The following examples introduce us to a number of other relationships.

Example 17 If $\sin\alpha = 3/5$, find the numerical value of $\cos\alpha$, $\tan\alpha$, and $\cot\alpha$.

Solution. The fraction 3/5 reminds us of our best friend, the 3-4-5 triangle:

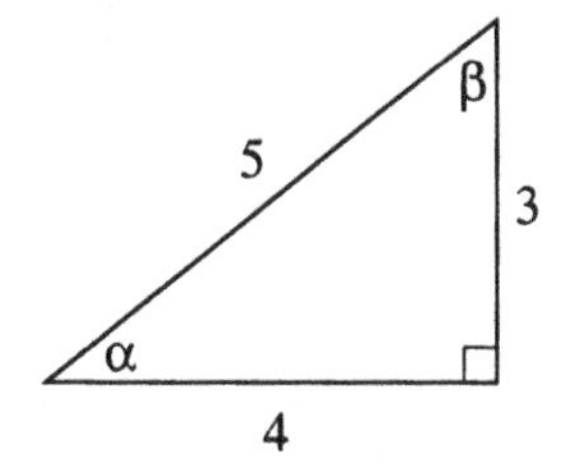

In fact, α is the measure of one of the angles in such a triangle: the one opposite the side of length 3 (see the diagram above). Having drawn this triangle, we easily see that $\cos\alpha = 4/5$, $\tan\alpha = 3/4$, and $\cot\alpha = 4/3$. □

Example 18 If $\sin\alpha = 2/5$, find the numerical value of $\cos\alpha$, $\tan\alpha$, and $\cot\alpha$.

Solution. We can again draw a right triangle with angle α:

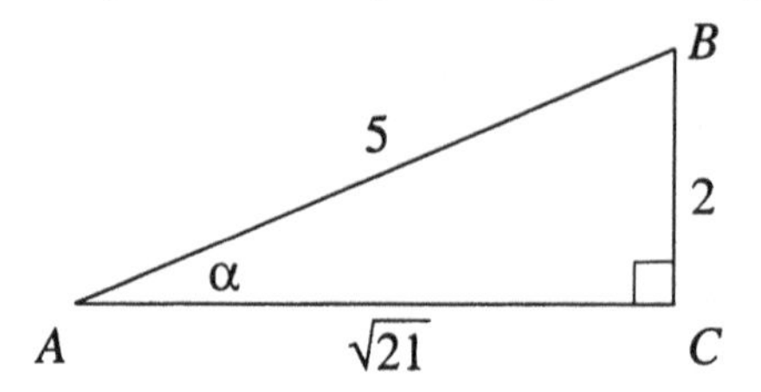

We know only two sides of the triangle. To find $\cos\alpha$, we need the third side. The Pythagorean theorem will give this to us. Because $a^2 + b^2 = c^2$, we have $b^2 + 2^2 = 5^2$, so $b^2 = 21$ and $b = \sqrt{21}$. Now we know all the sides of this triangle, and it is clear that

$$\cos\alpha = \frac{\sqrt{21}}{5}, \qquad \tan\alpha = \frac{2}{\sqrt{21}}, \qquad \cot\alpha = \frac{\sqrt{21}}{2}.$$

□

Example 19 We can find an answer to the question in Example 18 in a different way. For the same angle α, we have the right to draw a different right triangle, with hypotenuse 1, and leg 2/5. Do the calculation in this case for yourself. It will produce the same result. □

Example 20 If $\sin\alpha = a$, where $0 < a < 1$, express in terms of a the value of $\cos\alpha$, $\tan\alpha$, and $\cot\alpha$.

Solution. As before, we choose a right triangle with an acute angle equal to α:

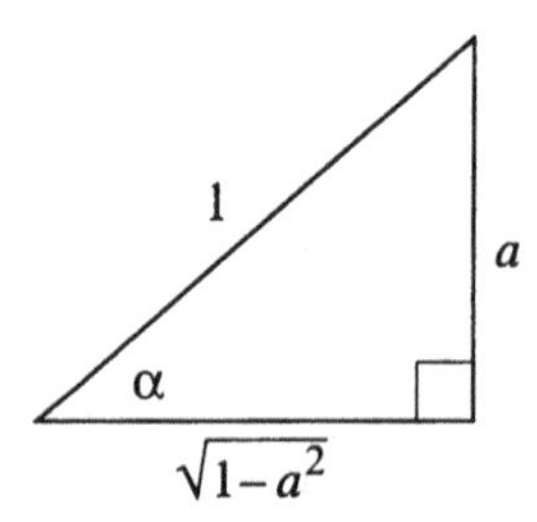

The simplest is one in which the hypotenuse has length 1 and the leg opposite α is a. Let the other leg be x. Then $x^2 + a^2 = 1$, so $x = \sqrt{1 - a^2}$. Now we know all the sides of this triangle, and we can write everything down easily:

$$\cos\alpha = \sqrt{1-a^2}, \qquad \tan\alpha = \frac{a}{\sqrt{1-a^2}}, \qquad \cot\alpha = \frac{\sqrt{1-a^2}}{a}. \qquad \square$$

Exercises

1. Suppose $\sin\alpha = 8/17$. Find the numerical value of $\cos\alpha$, $\tan\alpha$, and $\cot\alpha$.

2. Suppose $\cos\alpha = 3/7$. Find the numerical value of $\sin\alpha$, $\tan\alpha$, and $\cot\alpha$.

3. Suppose $\cos\alpha = b$. In terms of b, express $\sin\alpha$, $\tan\alpha$, and $\cot\alpha$.

4. Suppose $\tan\alpha = d$. In terms of d, express $\sin\alpha$, $\cos\alpha$, and $\cot\alpha$.

5. Fill in the following table. In each row, the value of one trigonometric function is assigned a variable. Express each of the other trigonometric functions in terms of that variable. The work for one of the rows is already done.

	$\sin\alpha$	$\cos\alpha$	$\tan\alpha$	$\cot\alpha$
$\sin\alpha$	a	$\sqrt{1-a^2}$	$\frac{a}{\sqrt{1-a^2}}$	$\frac{\sqrt{1-a^2}}{a}$
$\cos\alpha$		a		
$\tan\alpha$			a	
$\cot\alpha$				a

Please do not try to memorize this table. Its first row can be filled with the help of the triangle

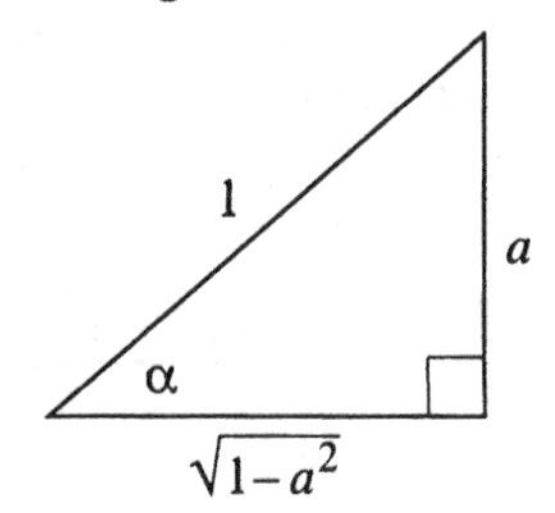

whose sides can be found from the Pythagorean theorem. For the other rows, you can use use the triangles

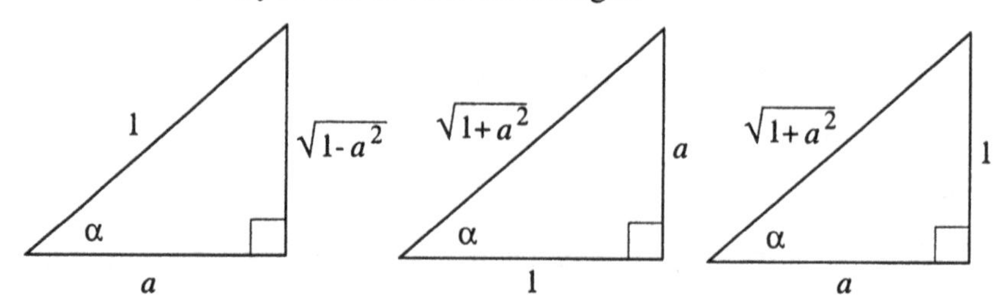

This is all you'll ever need.

Remark. We have implicitly assumed that for every number a between 0 and 1, there exists a right triangle that contains an angle whose sine is a. But this is clear from geometry: we can construct such a triangle by taking the hypotenuse to be 1, and one leg to be a.[1]

2 Algebra or geometry?

Example 21 Suppose $\sin\alpha = 1/2$. Find the numerical value of $\cos\alpha$, $\tan\alpha$, and $\cot\alpha$.

Solution. We can do this geometrically, by drawing a triangle (as in Exercise 5 above). Or we can do this algebraically, using the results of Example 20. For instance,

$$\cos\alpha = \sqrt{1-a^2} = \sqrt{1-\left(\frac{1}{2}\right)^2} = \sqrt{1-\frac{1}{4}} = \sqrt{\frac{3}{4}} = \frac{\sqrt{3}}{2}.$$

Or did you notice right away that α is an angle in one of our friendly triangles? □

[1] You may have some objection to taking the hypotenuse of our triangle to have length 1. If you insist, we can take some length c for this hypotenuse. We will then get the same results, but the calculations will be longer. For example, suppose $\sin\alpha = x$. Choose a right triangle that contains a, and that has hypotenuse of length c. Suppose the leg opposite α has length a. Then $a/c = x$, since $\sin\alpha = x$. So $a = cx$. If we are asked for $\cos\alpha$, we can suppose the length of the other leg is b. Then $b^2 = c^2 - a^2 = c^2 - c^2x^2$, and

$$\cos a = \frac{b}{c} = \frac{\sqrt{c^2-c^2x^2}}{c} = \frac{c\sqrt{1-x^2}}{c} = \sqrt{1-x^2}.$$

Exercises

1. Find $\sin^2 30°$.

   Solution. Since $\sin 30° = 1/2$, we see that

$$\sin^2 30° = \left(\frac{1}{2}\right)^2 = \frac{1}{4}. \qquad \square$$

2. Find $\sin^2 45°$. Check the result with a calculator.

3. Rewrite the table from Exercise 5 on page 43, but using the names of the trigonometric ratios. The first row below has been filled in as an example.

	$\sin\alpha$	$\cos\alpha$	$\tan\alpha$	$\cot\alpha$
$\sin\alpha$	$\sin\alpha$	$\sqrt{1-\sin^2\alpha}$	$\dfrac{\sin\alpha}{\sqrt{1-\sin^2\alpha}}$	$\dfrac{\sqrt{1-\sin^2\alpha}}{\sin\alpha}$
$\cos\alpha$				
$\tan\alpha$				
$\cot\alpha$				

3 A remark about names

We have already seen that if α and β are complementary, then $\sin\alpha = \cos\beta$. Historically, the prefix "co-" stands for "complement," since two acute angles in the same right triangle are complementary.

The following exercises extend this note.

Exercises

Using the diagram on the right, show that if α and β are complementary, then:

1. $\tan\alpha = \cot\beta$.
2. $\cot\alpha = \tan\beta$.
3. $\sec\alpha = \csc\beta$.
4. $\csc\alpha = \sec\beta$.

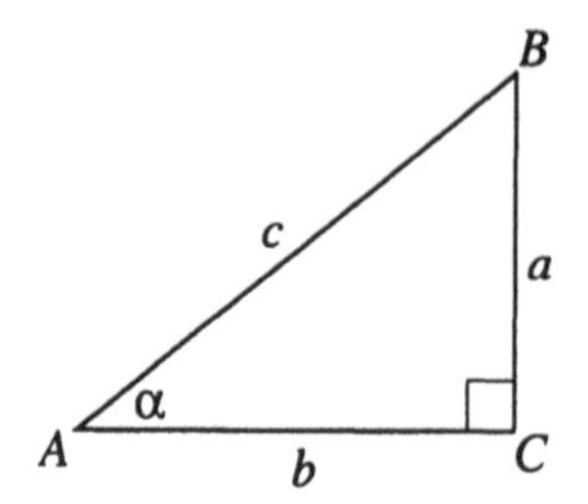

4 An identity crisis?

From the table on page 45 we see, for example, that $\cos\alpha = \sqrt{1-\sin^2\alpha}$. We have also seen that, for any angle α, $\sin^2\alpha + \cos^2\alpha = 1$. Such equations, which are true for every value of the variable, are called *identities*.

From the identities we have, we can derive many more. But there is no need for anxiety. We will not have an identity crisis. If you forget all these identities, they are easily available from the three fundamental identities below:

$$\sin^2\alpha + \cos^2\alpha = 1$$

$$\tan\alpha = \frac{\sin\alpha}{\cos\alpha}$$

$$\cot\alpha = \frac{1}{\tan\alpha}$$

From these simple identities we can derive many others involving the sine, cosine, tangent, and cotangent of a single angle. One way to derive a new identity is to draw a right triangle with an acute angle equal to α, and substitute a/c for $\sin\alpha$, b/c for $\cos\alpha$, and so on.

Example 22 Prove that $\tan\alpha = \dfrac{\sin\alpha}{\cos\alpha}$.

Solution. We can draw a right triangle with legs a, b, hypotenuse c, and acute angle α opposite the leg of length a. Then we have

$$\sin\alpha/\cos\alpha = \frac{(a/c)}{(b/c)} = (a/c)(c/b) = a/b = \tan\alpha\,.$$

□

Another way to prove a new identity is to show that it follows from other identities that we know already.

Example 23 Prove the identity $\tan\alpha\cot\alpha = 1$.

Solution. From our table, we see that $\tan\alpha = \sin\alpha/\cos\alpha$. We also see that $\cot\alpha = \cos\alpha/\sin\alpha$. Therefore,

$$\tan\alpha\cot\alpha = \left(\frac{\sin\alpha}{\cos\alpha}\right)\left(\frac{\cos\alpha}{\sin\alpha}\right) = 1\,. \qquad \square$$

Example 24 Show that $\tan^2\alpha + 1 = 1/\cos^2\alpha$.

Solution. We know that $\sin^2\alpha + \cos^2\alpha = 1$, so

$$\frac{\sin^2\alpha}{\cos^2\alpha} + \frac{\cos^2\alpha}{\cos^2\alpha} = \frac{1}{\cos^2\alpha}\,,$$

or

$$\tan^2\alpha + 1 = \frac{1}{\cos^2\alpha}\,. \qquad \square$$

You will have a chance to practice both these techniques in the exercises below.

Exercises

1. Verify that $\sin^2\alpha + \cos^2\alpha = 1$ if α equals 30°, 45°, and 60°.
2. The sine of an angle is $\sqrt{5}/4$. Express in radical form the cosine of this angle.
3. The cosine of an angle is $2/3$. Express in radical form the sine of the angle.
4. The tangent of an angle is $1/\sqrt{3}$. Find the numerical value of the sine and cosine of this angle.
5. Prove the following identities for an acute angle α:

   a) $\cot x \sin x = \cos x$.

b) $\dfrac{\tan x}{\sin x} = \dfrac{1}{\cos x}$.

c) $\cos^2\alpha - \sin^2\alpha = 2\cos^2\alpha - 1$.

d) $\dfrac{\sin\alpha}{1+\cos\alpha} = \dfrac{1-\cos\alpha}{\sin\alpha}$.

e) $(\sin^2\alpha + 2\cos^2\alpha - 1)/\cot^2\alpha = \sin^2\alpha$.

f) $\cos^2\alpha = 1/(1+\tan^2\alpha)$.

g) $\sin^2\alpha = 1/(\cot^2\alpha + 1)$.

h) $\dfrac{1-\cos\alpha}{1+\cos\alpha} = \left(\dfrac{\sin\alpha}{1+\cos\alpha}\right)^2$.

i) $\dfrac{\sin^3\alpha - \cos^3\alpha}{\sin\alpha - \cos\alpha} = 1 + \sin\alpha\cos\alpha$.

6. a) For which angles α is $\sin^4\alpha - \cos^4\alpha > \sin^2\alpha - \cos^2\alpha$?

   b) For which angles α is $\sin^4\alpha - \cos^4\alpha \geq \sin^2\alpha - \cos^2\alpha$?

7. If $\tan\alpha = 2/5$, find the numerical value of $2\sin\alpha\cos\alpha$.

8. a) If $\tan\alpha = 2/5$, find the numerical value of $\cos^2\alpha - \sin^2\alpha$.

   b) If $\tan\alpha = r$, write an expression in terms of r that represents the value of $\cos^2\alpha - \sin^2\alpha$.

9. If $\tan\alpha = 2/5$, find the numerical value of $\dfrac{\sin\alpha - 2\cos\alpha}{\cos\alpha - 3\sin\alpha}$.

10. If $\tan\alpha = 2/5$, and a, b, c, d are arbitrary rational numbers, with $5c + 2d \neq 0$, show that $\dfrac{a\sin\alpha + b\cos\alpha}{c\cos\alpha + d\sin\alpha}$ is a rational number.

11. For what value of α is the value of the expression $(\sin\alpha + \cos\alpha)^2 + (\sin\alpha - \cos\alpha)^2$ as large as possible?

5 Identities with secant and cosecant

While we do not often have to use the secant and cosecant, it is often convenient to express the fundamental identities above in terms of these two ratios. We can always restate the results as desired, using the fact that $\sec\alpha = 1/\cos\alpha$ and $\csc\alpha = 1/\sin\alpha$.

Example 25 Show that $\sec^2\alpha = 1 + \tan^2\alpha$ for any acute angle α.

Solution. We know that

$$\sec\alpha = \frac{1}{\cos\alpha},$$

so the given identity is equivalent to the statement that

$$\frac{1}{\cos^2\alpha} = 1 + \tan^2\alpha .$$

This last identity was proven in Example 24, page 47. □

Exercises

1. Rewrite each given identity using only sine, cosine, tangent or cotangent.

   a) $\tan\alpha\csc\alpha = \sec\alpha$.

   b) $\cot\alpha\sec\alpha = \csc\alpha$.

   c) $\dfrac{1}{\sec\alpha}\csc\alpha = \cot\alpha$.

   d) $\tan^2\alpha = (\sec\alpha + 1)(\sec\alpha - 1)$.

   e) $\csc^2\alpha = 1 + \cot^2\alpha$.

2. Rewrite in terms of secant and cosecant, tangent or cotangent. Simplify your answers so that they do not involve fractions.

   a) $\dfrac{\tan\alpha}{\sin\alpha} = \dfrac{1}{\cos\alpha}$.

   b) $\dfrac{1}{\sin\alpha}\cos\alpha = \cot\alpha$.

   c) $\tan^2\alpha + 1 = \dfrac{1}{\cos^2\alpha}$.

   d) $\dfrac{1}{\sin^2\alpha} = 1 + \cot^2\alpha$.

6 A lemma

We have already seen that if $a = \cos\alpha$ and $b = \sin\alpha$ for some acute angle α, then $a^2 + b^2 = 1$. We can also prove the converse of this statement: If a and b are some pair of positive numbers such that $a^2 + b^2 = 1$, then there exists an angle θ such that $a = \cos\theta$ and $b = \sin\theta$. Indeed, if we draw a triangle with sides a, b, and 1, the Pythagorean theorem (statement II) guarantees us that this is a right triangle. Then the angle θ that we are looking for appears in the triangle "automatically."

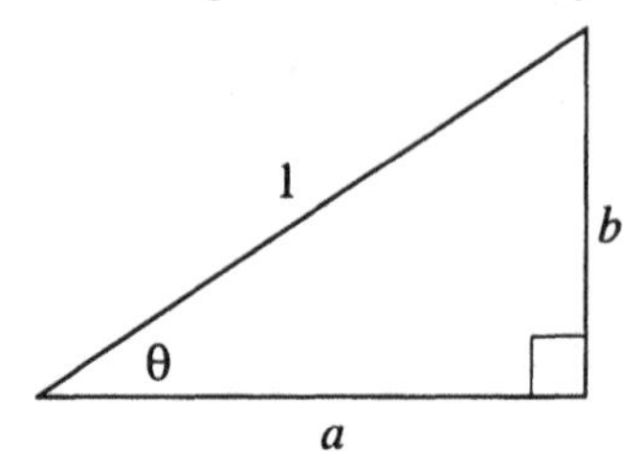

Exercises

1. Suppose α is some angle less than 45°. If $a = \cos^2\alpha - \sin^2\alpha$ and $b = 2\sin\alpha\cos\alpha$, show that there is an angle θ such that $a = \cos\theta$ and $b = \sin\theta$.

2. Suppose that α is some angle. If $a = \sqrt{(1+\cos\alpha)/2}$ and $b = \sqrt{(1-\cos\alpha)/2}$, show that there is an angle θ such that $a = \cos\theta$ and $b = \sin\theta$.

3. Suppose that α is some angle. If $a = 4\cos^3\alpha - 3\cos\alpha$ and $b = 3\sin\alpha - 4\sin^3\alpha$, show that there is an angle θ such that $a = \cos\theta$ and $b = \sin\theta$.

4. Suppose that t is a number between 0 and 1. If
$$a = \frac{1-t^2}{1+t^2} \quad \text{and} \quad b = \frac{2t}{1+t^2},$$
show that there is an angle θ such that
$$a = \cos\theta \quad \text{and} \quad b = \sin\theta\,.$$

5. (A non-trigonometric identity) If $p^2 + q^2 = 1$, show that $(p^2 - q^2)^2 + (2pq)^2 = 1$ also. Which trigonometric identity, of those in the exercises above, is this similar to?

7 Some inequalities

You may remember from geometry that the hypotenuse is the largest side in a right triangle (since it is opposite the largest angle). So the ratio of any leg to the hypotenuse of a right triangle is less than 1. It follows that $\sin\alpha < 1$ and $\cos\alpha < 1$ for any acute angle α.

That is all the background you need to do the following exercises.

Exercises

1. For any acute angle α, show that $1 - \sin\alpha \geq 0$. For what value(s) of α do we have equality?

2. For any acute angle α, show that $1 - \cos\alpha \geq 0$. For what value(s) of α do we have equality?

3. Which of the following statements are true for all values of α?

   a) $\sin^2\alpha + \cos^2\alpha = 1$.
   b) $\sin^2\alpha + \cos^2\alpha \geq 1$.
   c) $\sin^2\alpha + \cos^2\alpha \leq 1$.

   Answer. They are all correct. Can you see why?

4. There are 4 supermarkets having a sale. Which of these are offering the same terms for their merchandise?

   - In supermarket A, everything costs no more than $1.
   - In supermarket B, everything costs less than $1.
   - In supermarket C, everything costs $1 or less.
   - In supermarket D, everything costs more than $1.

5. Which inequality is correct?

   a) For any angle α, $\sin\alpha + \cos\alpha < 2$.
   b) For any angle α, $\sin\alpha + \cos\alpha \leq 2$.

6. What is the largest possible value of $\sin\alpha$? Of $\cos\alpha$?

8 Calculators and tables

It is, in general, very difficult to get the numerical value of the sine of an angle given its degree measure. For example, how can we calculate $\sin 19°$?

One way would be to draw a right triangle with a 19° angle, and measure its sides very accurately. Then the ratio of the side opposite the angle to the hypotenuse will be the sine of 19°.

But this is not a method that mathematicians like. For one thing, it depends on the accuracy of our diagram, and of our rulers. We would like to find a way to calculate $\sin 19°$ using only arithmetic operations. Over the centuries, mathematicians have devised some very clever ways to calculate sines, cosines, and tangents of any angle without drawing triangles.

We can benefit from their labors by using a calculator. Your scientific calculator probably has a button labeled "sin," another labeled "cos," and a third labeled "tan." These give approximate values of the sine, cosine, and tangent (respectively) of various angles.

Warning: Most "nice" angles do not have nice values for sine, cosine, or tangent. The values of $\tan 61°$ or $\sin 47°$ will not be rational, and will not even be a square root or cube root of a rational number. There are a very few angles with integer degree measures and "nice" values for sine, cosine, or tangent.

Exercises

1. Find a handheld scientific calculator, and get from it the values of $\sin 30°$, $\sin 45°$, and $\sin 60°$. Compare these values with those we found in Chapter 1.

2. Betty thinks that the tangent of 60° is $\sqrt{3}$. How would you check this using a scientific calculator?

3. How would you use a calculator to get the cotangent of an angle of 30°? of 20°?

   Hint: Many calculators have a button labeled "1/x." If you press this, the display shows the reciprocal of the number previously displayed.

4. Fill in the following tables:

in radical or rational form				
α	$\sin\alpha$	$\cos\alpha$	$\tan\alpha$	$\cot\alpha$
30°				
45°				
60°				

in decimal form, from calculator				
α	$\sin\alpha$	$\cos\alpha$	$\tan\alpha$	$\cot\alpha$
30°				
45°				
60°				

9 Getting the degree measure of an angle from its sine

Example 26 What is the degree measure of the smaller acute angle of a right triangle with sides 3, 4, and 5?

Solution. We could draw a very accurate diagram, and use a very accurate protractor to answer this question. But again, mathematicians have developed methods that do not depend on the accuracy of our instruments. Your calculator uses these methods, but you must know how the buttons work.

The sine of the angle we want is $3/5 = .6$. Enter the number $.6$, then look for the button marked "arcsin" or "$\sin^{-1}$" (for some calculators, you must press this button first, then enter .6). You will find that pushing this button gives a number close to 36°. This is the angle whose sine is $.6$. □

On a calculator, you can read the symbol "arcsin" or "$\sin^{-1}$" as "the angle whose sine is ... " Similarly, "arccos" means "the angle whose cosine is ... " and "$\tan^{-1}$" means "the angle whose tangent is ... "

Exercises

1. In the text, we found an estimate for the degree-measure of the smaller acute angle in a 3-4-5 triangle. Using your calculator, find, to the nearest degree, the measure of the larger angle. Using your estimate, does the sum of the angles of such a triangle equal 180 degrees?

2. Using your calculator, find

   a) arcsin 1.

   b) arccos 0.7071067811865.

3. Using your calculator, find the angle whose cosine is .8.

4. Using your calculator, find the angle whose sine is .6.

5. We know that $\sin 30° = .5$. Write down your estimate for $\sin 15°$, then check your estimate with the value from a table or calculator.

6. Suppose $\sin x = .3$. Use your calculator to get the degree-measure of x. Now check your answer by taking the sine of the angle you found.

7. Suppose $\arcsin x = 53°$. Use your calculator to get an estimate for the value of x. Now check your answer by taking the arcsin of the number you found.

8. If $\arcsin x = 60°$, find x, without using a calculator.

9. Using your calculator, find arcsin (sin 17°).

10. Using your calculator, find sin (arcsin 0.4).

11. Find arcsin (sin 30°) without using your calculator. Then find sin (arcsin 1/2), without using the calculator. Explain your results.

12. With a calculator, check that $\cos^2 A + \sin^2 A = 1$ if A equals 20° and if A equals 80°.

13. With a calculator, check that $\tan A = \sin A / \cos A$ if A equals 20° and if A equals 80°.

14. Using a calculator to get numerical values, draw a graph of the value of $\sin x$ as x varies from 0° to 90°.

10 Solving right triangles

Many situations in life call for the solution of problems like the following.

Example 27 The hypotenuse of a right triangle is 5, and one of its acute angles is 37 degrees. Find the other two sides.

Solution. From a calculator, we obtain $\sin 37° \approx 0.6018$ and $\cos 37° \approx 0.7986$.

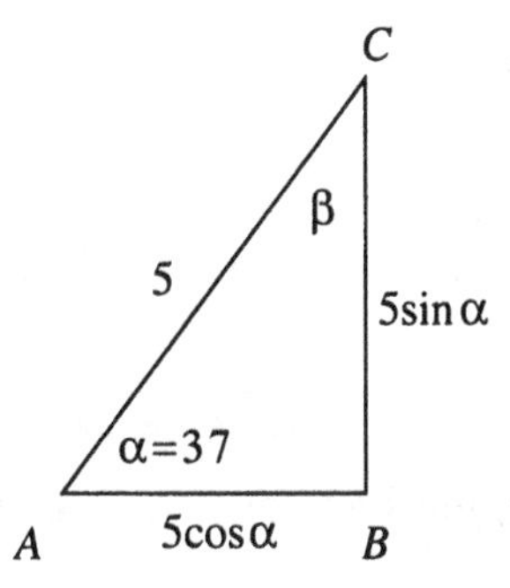

Since $\sin\alpha = BC/AC = BC/5$, we have that $BC = 5\sin 37° = 5 \times 0.6018 = 3.009$. Similarly, $AC = 5\cos 37° = 5 \times 0.7986 = 3.993$. Both values are correct to the nearest thousandth. □

Exercises

1. Find the legs of a right triangle with hypotenuse 9 and an acute angle of 72 degrees.

2. The two legs of a right triangle are 7 and 10. Find the hypotenuse and the two acute angles.

3. A right triangle has a leg of length 12. If the acute angle opposite this leg measures 27 degrees, find the other leg, the other acute angle, and the hypotenuse.

4. A right triangle has a leg of length 20. If the acute angle adjacent to this leg measures 73 degrees, find the other leg, the other acute angle, and the hypotenuse.

11 Shadows

Because the sun is so far away from the earth, the rays of light that reach it from the earth are almost parallel. If we think of a small area of the earth as flat (and we usually do!), then the sun's rays strike this small region at the same angle:

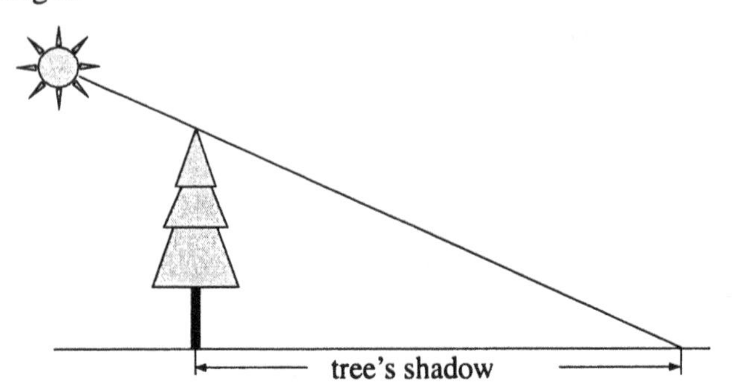

So we can, for example, tell how long the shadow of an object will be, given its length.

Example 28 If the rays of the sun make a 23° angle with the ground, how long will the shadow be of a tree which is 20 feet high?

Solution. In the diagram below, AC is the tree, and BC is the shadow:

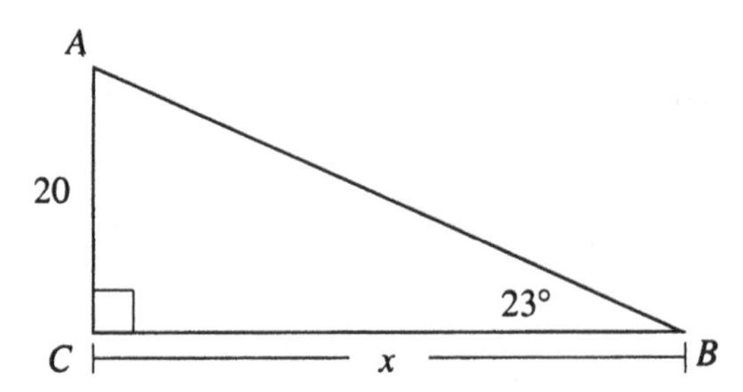

We have that

$$\tan 23^\circ = \frac{AC}{BC} = \frac{20}{x}, \text{ or } x = \frac{20}{\tan 23^\circ}.$$

From a calculator, $\tan 23^\circ \approx 0.4244$, so $x \approx 20/0.4244 \approx 47.13$ feet. □

Exercises

1. When the sun's rays make an angle with the ground of 46 degrees, how long is the shadow cast by a building 50 feet high?

2. At a certain moment, the sun's rays strike the earth at an angle of 32 degrees. At that moment, a flagpole casts a shadow which is 35 feet long. How tall is the flagpole?

3. Why are shadows longest in the morning and evening? When would you expect the length of a shadow to be the shortest?

4. Can it happen that an object will not cast any shadow at all? When and where? You may need to know something about astronomy to investigate this question.

12 Another approach to the sine ratio

There is a simple connection between the sine of an angle and chords in a circle.

Theorem If α is the angle subtended by a chord PB at a point on a circle of radius r (such as point A in the diagram below), then

$$\sin\alpha = \frac{PB}{2r}.$$

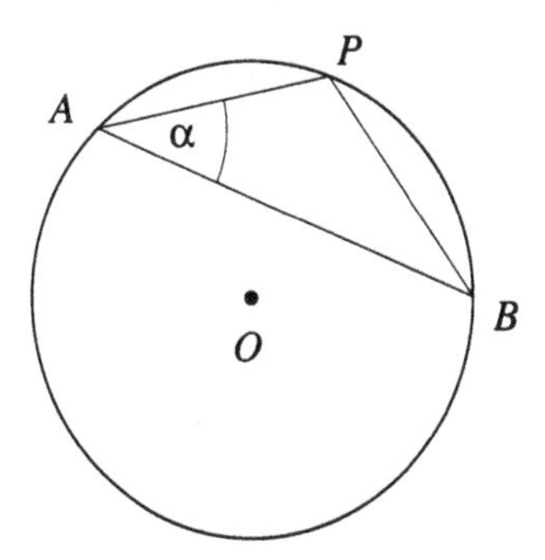

Before we prove this theorem, let us resolve a problem in the way it is stated. We can pick different points on the circle (such as A' and A'' in the figure below), and consider the different angles subtended by the same chord PB:

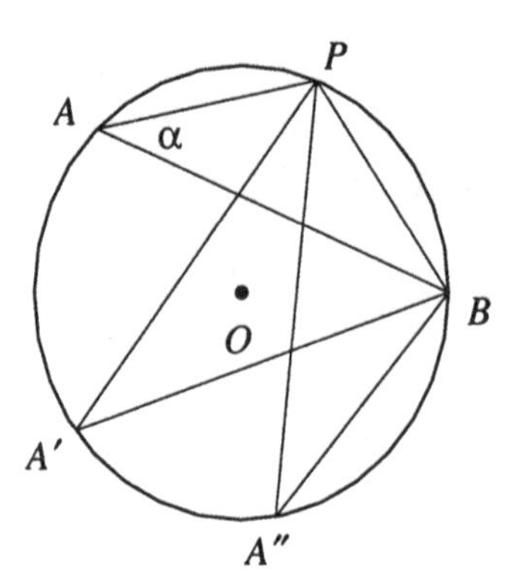

Does it matter which point we pick? No, it does not. An important theorem of geometry asserts that all the angles subtended by a given chord in a circle are equal[2], so that it makes no difference which point on (major) arc PB we choose.

For this reason, we can prove our theorem by making a very special choice: for point A, we choose the point diametrically opposite to point B:

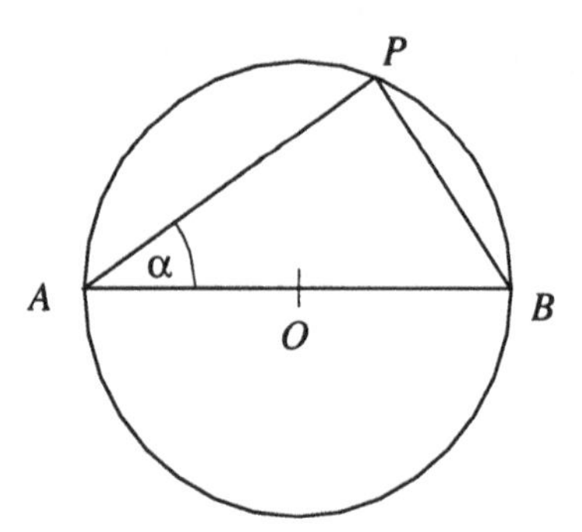

The same geometric theorem about inscribed angles assures us that $\angle APB$ is a right angle, so $\sin\alpha = PB/AB = PB/2r$. This completes the proof.

We can give another proof of this theorem. Consider the diagram

[2]See the theorem on inscribed angles in the appendix to this chapter.

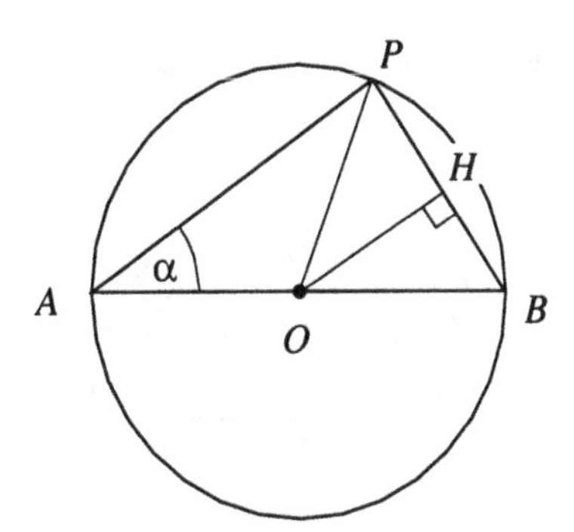

We have shown here a perpendicular from the center of the circle O to the chord PB. Geometry tell us that the foot of this perpendicular H is the midpoint of PB.

Then, as before, $\angle APB = 90°$, which implies that AP and HO are parallel. Thus, $\angle HOB = \angle PAB = \alpha$. Now in right triangle HOB, we have

$$\sin\alpha = \frac{HB}{OB} = \frac{1}{2}\cdot\frac{PB}{OB} = \frac{PB}{2r},$$

as we know from the first proof.

Exercises

1. The diagram on the left below shows a chord AB and its central angle $\angle AOB = \theta$:

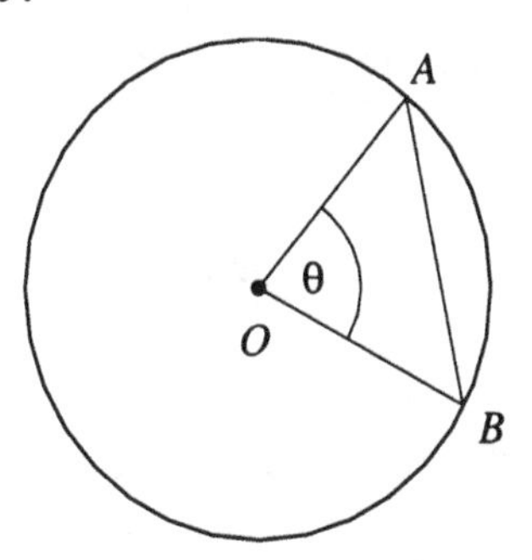

Suppose the diameter of the circle is 1. How is the length of AB related to θ?

Answer. $AB = \sin(\theta/2)$.

2. Now, using the same diagram, suppose the *radius* of the circle is 1. How is the length of AB related to θ now?

Answer. $AB = 2\sin(\theta/2)$.

3. The diagram below shows a circle of diameter 1, and two acute angles θ and φ:

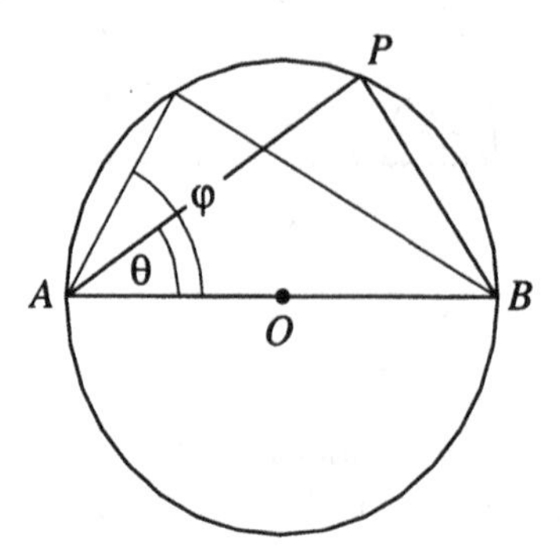

How does the diagram suggest that if $\varphi > \theta$, then $\sin\varphi > \sin\theta$?

4. We know from geometry that a circle may be drawn through the three vertices of any triangle. Find the radius of such a circle if the sides of the triangle are 6, 8, and 10.

5. Starting with an acute triangle, we can draw its circumscribed circle (the circle that passes through its three vertices). If α is any one of the angles of the triangle, show that the ratio $a : \sin\alpha$ is equal to the diameter of the circle.

6. Use Exercise 5 to show that if α, β, γ are three angles of an acute triangle, and a, b, and c are the sides opposite them respectively, then $\dfrac{a}{\sin\alpha} = \dfrac{b}{\sin\beta} = \dfrac{c}{\sin\gamma}$.

7. The diagram below shows a circle with center O, and chords AB and AC:

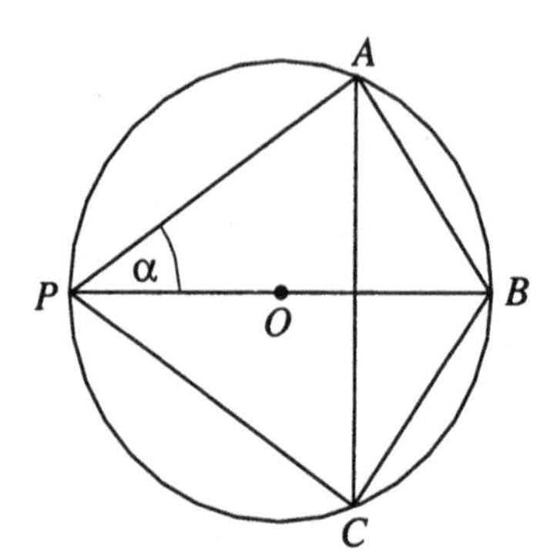

Arc AC is double arc AB. Diameter BP, chord AP and chord CP are drawn in, and $BP = 1$ (the diameter of the circle has unit length). If angle APB measures α degrees, use this diagram to show that $\sin 2\alpha < 2 \sin \alpha$.

You may need the theorem known as the *triangle inequality*: The sum of the lengths of any two sides of a triangle is greater than the length of the third side.

8. In a circle of diameter 10 units, how long is a chord intercepted by an inscribed angle of 60 degrees?

9. In a circle of diameter 10 units, how long is a chord intercepted by a central angle of 60 degrees?

10. Find the length of a side of a square inscribed in a circle of diameter 10 units.

11. If you knew the exact numerical value of $\sin 36°$, how could you calculate the side of a regular pentagon inscribed in a circle of diameter 10?

Appendix – Review of Geometry

I. Measuring arcs

One natural way to measure an arc of a circle is to ask what portion of its circle the arc covers. We can look at the arc from the point of view of the center of the circle, and draw the central angle that cuts off the arc:

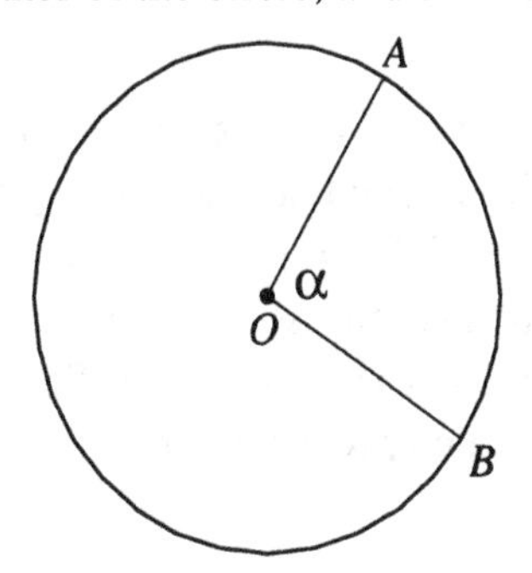

If central angle AOB measures α degrees, then we say that arc $\overset{\frown}{AB}$ measures α degrees as well.

Exercises

1. What is the degree measure of a semicircle? A quarter of a circle?

2. What is the degree measure of the arc cut off by one side of a regular pentagon inscribed in a circle? A regular hexagon? A regular octagon?

II. Inscribed angles and their arcs

An important theorem of geometry relates the degree-measure of an arc not to its central angle, but to any *inscribed* angle which intercepts that arc:

Theorem The degree measure of an inscribed angle is half the degree measure of its intercept arc.

Proof We divide the proof into three cases.

1: First we prove the statement for the case in which one side of the inscribed angle is a diameter.

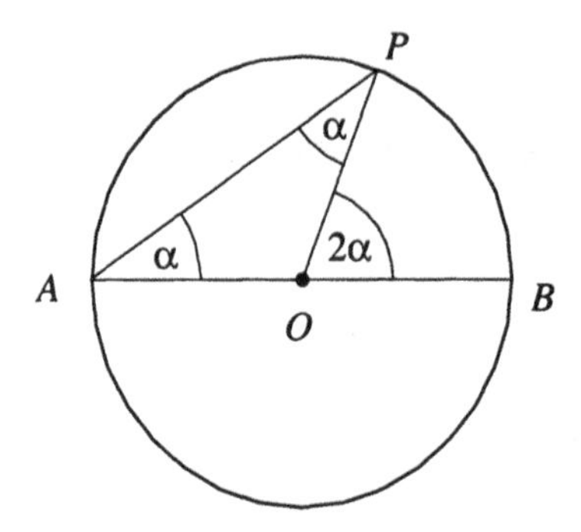

Take inscribed angle PAB, and draw PO (where O is the center of the circle). Since OP and OA are radii of the circle, they are equal, and triangle PAO is isosceles. Hence $\angle APO = \angle PAO = \alpha$. But $\angle POB$ is an exterior angle of this triangle, and so is equal to the sum of the remote interior angles, which is $\alpha + \alpha = 2\alpha$. So the degree-measure of arc $\overset{\frown}{PB}$ is also 2α, which proves the theorem for this case.

2: Suppose the center of the circle is not on one side of the inscribed angle, but inside it.

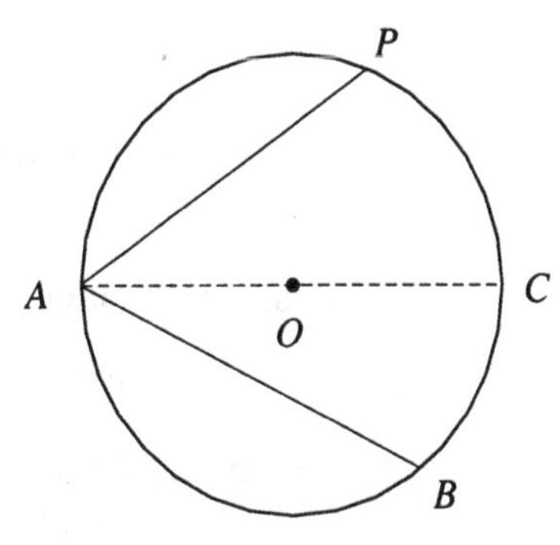

If we draw diameter AC, then angles PAC, BAC are inscribed angles covered by Case 1, so $\angle PAC = \frac{1}{2}\widehat{PC}$ and $\angle CAB = \frac{1}{2}\widehat{CB}$. Now $\angle PAB = \angle PAC + \angle CAB = \frac{1}{2}\widehat{PC} + \frac{1}{2}\widehat{CB} = \frac{1}{2}\widehat{PB}$, which is what we wanted to prove.

3: Suppose the center of the circle is outside the inscribed angle.

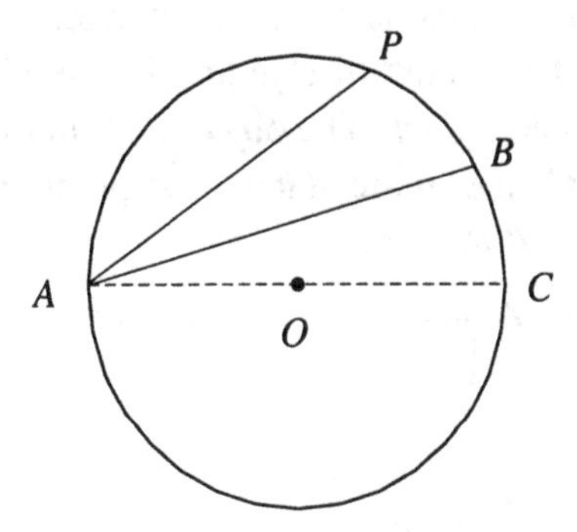

If we draw diameter AC, then angles PAC, BAC are inscribed angles covered by Case 1, so $\angle PAC = \frac{1}{2}\widehat{PC}$ and $\angle CAB = \frac{1}{2}\widehat{CB}$. Now $\angle PAB = \angle PAC - \angle CAB = \frac{1}{2}\widehat{PC} - \frac{1}{2}\widehat{CB} = \frac{1}{2}\widehat{PB}$, which is what we had to prove. □

As a corollary to the theorem above, we state Thales's theorem, one of the oldest mathematical results on record:

Theorem An angle inscribed in a semicircle is a right angle.

The proof is a simple application of the previous result, and is left for the reader as an exercise.

Exercises

1. If two inscribed angles intercept the same arc, show that they must be equal.
2. Find the degree-measure of an angle of a regular pentagon.

   Hint: Any regular pentagon can be inscribed in a circle.
3. If a quadrilateral is inscribed in a circle, show that its opposite angles must be supplementary.

III. ... and conversely

If we have a particular object, which we will represent as a line segment, we are sometimes not so much interested in how big it *is*, but how big it *looks*. We can measure this by seeing how much of our field of vision the object takes up. If we think of standing in one place and looking all around, our field of vision is 2π. The object (AB in the diagram below) is *seen* at the angle APB if you are standing at point P. We often say that AB *subtends* angle APB at point P.

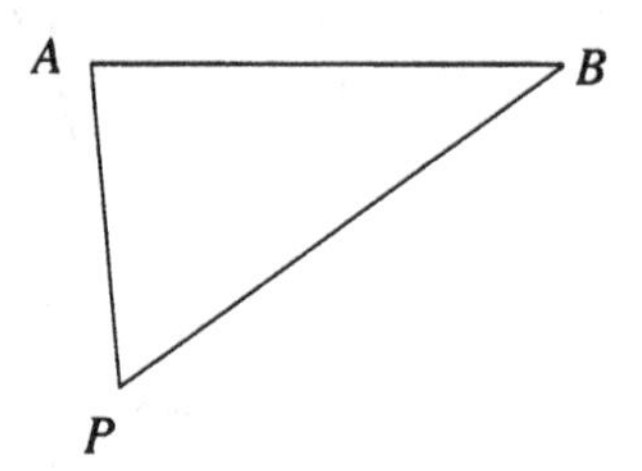

For example, viewed from the earth, the angle subtended by a star is very, very small, although we know that the star is actually very large. And the angle subtended by the sun is much greater, although we know that the sun, itself a star, is not the largest one.

Suppose the angle subtended by object AB at P measures 60°. Can we find other points at which AB subtends the same angle? From what positions does it subtend a larger angle? From what positions a smaller angle?

The answer is interesting and important. If we draw a circle through points A, B and P, then AB will subtend a 60° angle at any point on the circle, to one side of line AB:

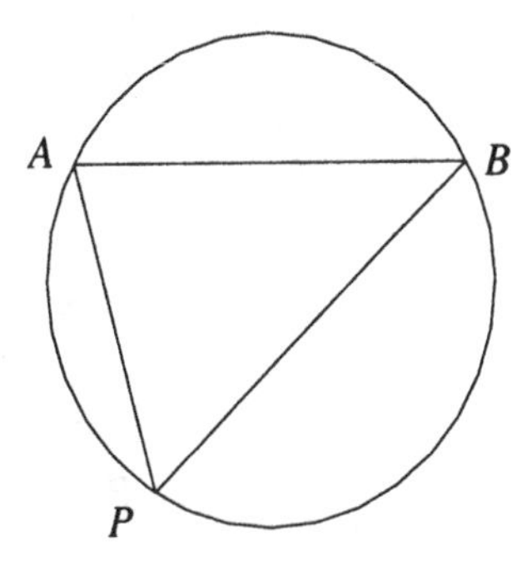

Also, AB will subtend an angle greater than 60° at any point inside the circle (to one side of line AB), and will subtend an angle less than 60° at any point outside the circle.

All this follows from the converse theorem to the one in the previous section:

Theorem Let AB subtend a given angle at some point P. Choose another point Q on the same side of the line AB as point P. Then

- If AB subtends the same angle at point Q as at point P, then Q is on the circle through A, B and P.
- If AB subtends a greater angle at Q, then Q is inside the circle through A, P and B.
- If AB subtends a smaller angle at Q, then Q is outside the circle through A, P and B.

(Remember that Q and P must be on the same side of line AB.) The proof of this converse will emerge from the exercises below.

Exercises

1. From what points will the object AB subtend an angle of 120°?
2. From what points will the object AB subtend an angle of 90°?
3. The diagrams below show an object AB, which subtends angle α at point P. Using these diagrams below, show that if point Q is outside the circle, then AB subtends an angle less than α at point Q, and if point R is inside the circle, AB subtends an angle greater than α at point R.

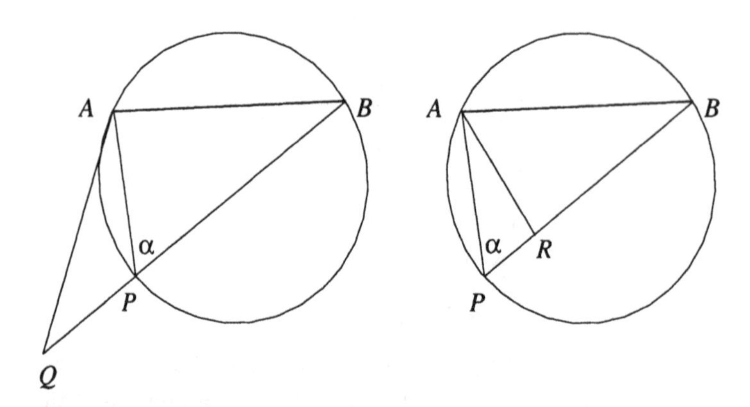

4. How does Exercise 3 prove that if an angle is half of a given arc, then it is inscribed in the circle of that arc?

5. The set of points P at which object AB subtends an angle equal to α is not the whole circle, but only the arc APB. What angle does AB subtend from the other points on the circle?

Chapter 3

Relationships in a Triangle

1 Geometry of the triangle

We would like to develop some applications of the trigonometry we've learned to geometric situations involving a triangle.

Let us work with the three sides and three angles of the triangle.

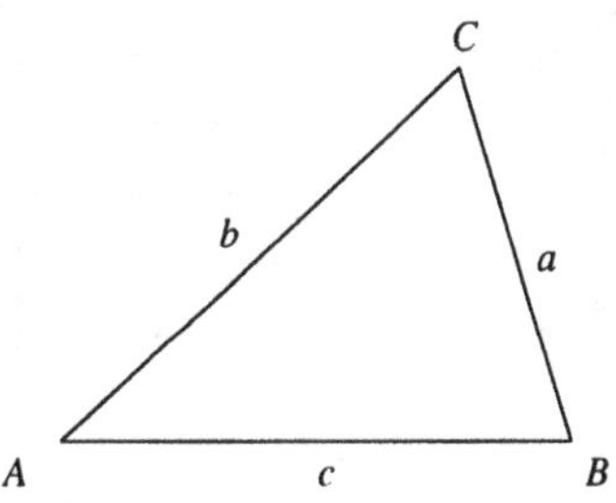

How many of these measurements do we need in order to reconstruct the triangle?[1]

This question is the subject of various "congruence theorems" in geometry. For example, if we know a, b and c (the three sides), the "SSS theorem" tells us that the three angles are determined. Any two triangles with the same three side-lengths are congruent.

But can we use any three side-lengths we like to make up our triangle? The "triangle inequality" of geometry tells us no. We must be sure that the

[1] Remember that if points A, B, and C are the vertices of a triangle, then we will also call the measures of the angles of the triangle A, B, and C. Then the length of the sides opposite angles A, B, and C are called a, b, and c, respectively. There are also other "parts" of a triangle: its area, angle bisectors, altitudes, medians, and still more interesting lines and measurements.

sum of any two of our three lengths is greater than the third; otherwise, the sides don't make a triangle. With this restriction, we can say that the three side-lengths of a triangle determine the triangle.

What other sets of measurements can determine a triangle? A little reflection will show that we will always need at least three parts (sides or angles), and various theorems from geometry will help us in answering this question.

Exercise

1. The table below gives several sets of data about a triangle. For example, "ABa" means that we are discussing two angles and the side opposite one of these angles. Some of the cases listed below are actually duplicates of others.

	Data	Determine a triangle?	Restrictions?
1	ABa		
2	ABb		
3	ABc		
4	AbC		
5	ABC		
6	Abc		
7	Bbc		
8	Cbc		

For each case, decide whether the given data determines a triangle. What restrictions must we place on the data so that a triangle can be formed? Some of these restrictions are a bit tricky. The case "*abc*" was discussed above.

Please do not memorize this table! We just want you to recall what geometry tells us – and what it does not tell us – about a triangle.

2 The congruence theorems and trigonometry

Some of the sets of data described above determine a triangle. For example, "SAS" data (the lengths of two sides of a triangle and the measure of the angle between them) always determines a triangle, and there is always a triangle which three such measurements specify.

But suppose we are given the lengths of two sides and the included angle in a triangle. How can we compute the length of the third side, or the degree-measures of the other two angles? The SAS statement of geometry doesn't tell us this. The next series of results will allow us to find missing parts of a triangle in this situation, and also in many others.

3 Sines and altitudes

A triangle has six basic elements: the three sides and the three angles. We would like to explore the relationship between these six basic elements and other elements of a triangle.

We begin with altitudes

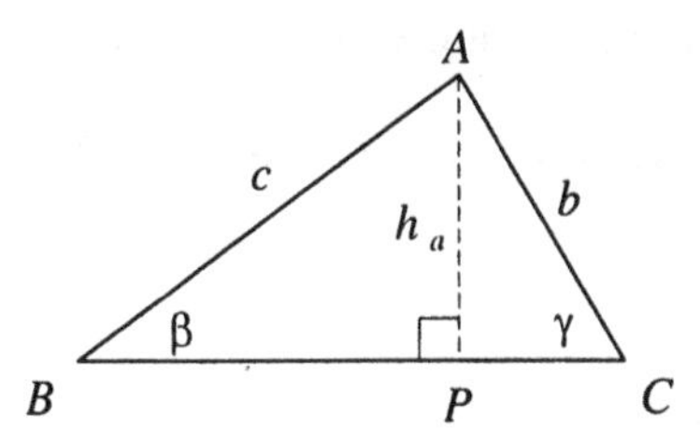

The diagram shows triangle ABC and the altitude to side BC. We use the symbol h_a to denote the length of this altitude, since it is the height to side a in the triangle. Similarly, we use h_b and h_c to denote the altitudes to sides b and c, respectively.

We can use the sine ratio to express h_a in terms of our six basic elements. In fact, we can do this in two different ways: from right triangle ABP, we have $\sin\beta = h_a/c$, so that

$$h_a = c\sin\beta\,.$$

From right triangle ACP, we have $\sin\gamma = h_a/b$, so

$$h_a = b\sin\gamma\,.$$

We can get formulas for each of the other altitudes by replacing each side with another side and the corresponding angles. This replacement is made easier if we think of it as a "cyclic" substitution. That is, we replace:

i) a with b, b with c, and c with a, and

ii) α with β, β with γ, and γ with α.

We obtain the following two new sets of relations:

$$\begin{aligned} h_b &= c \sin \alpha &= a \sin \gamma \,, \\ h_c &= b \sin \alpha &= a \sin \beta \,. \end{aligned}$$

Exercises

1. By drawing diagrams showing h_b and h_c, check that these last two sets of relations are correct.

2. In triangle ABC, $\alpha = 70°$ and $b = 12$. Find h_c.

3. Check to see that the expressions for the altitudes of a triangle are correct when the triangle is right-angled. (Remember how we defined the sine of a right angle on page 32.)

4. In triangle PQR, $p = 10$, $q = 12$ and $\angle PRQ = 30°$. Find its area.

4 Obtuse triangles

If a triangle contains an obtuse angle, two of its altitudes will fall outside the triangle. In the section above, we have not taken this possibility into account. Let us now correct this oversight.

In triangle ABC below, angle β is obtuse. Let us again try to express its altitude h_a in terms of its basic elements (sides and angles). (Note that h_a lies outside the triangle.)

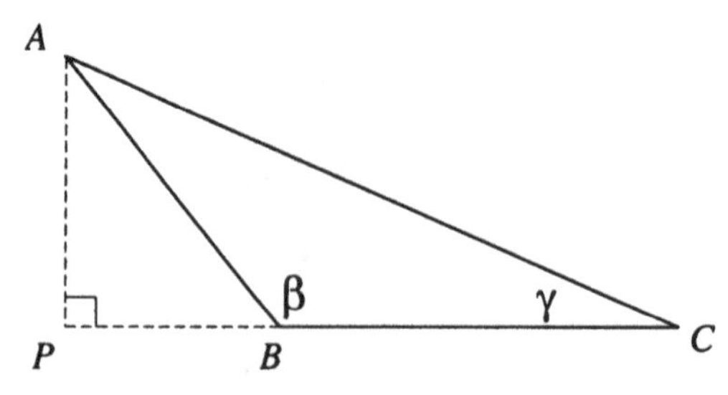

As before, triangle APC is a right triangle, so we have

$$AP = h_a = b \sin \gamma \,.$$

By the result in the previous section, we would also expect that

$$AP = h_a = c \sin \beta .$$

But we have no definition for $\sin \beta$, since β is an obtuse angle.

We can remedy the situation by looking at right triangle APB, in which $AB = c$. We find that $AP = AB \times \sin (\angle ABP)$, or

$$h_a = c \sin (180^\circ - \beta) .$$

This formula is a little bit cumbersome, so we take a rather daring step. We *define* $\sin \beta$ to be the same as $\sin (180^\circ - \beta)$.

In fact, we make the following general agreement:

Definition The sine of an obtuse angle is equal to the sine of its supplement.

Then we can write $h_a = c \sin \beta$ even when β is an obtuse angle. The remaining relations in such a case will follow from our rule for cyclic substitution, which still holds.

As we will see, this definition is convenient, not just to obtain this formula, but for other applications of trigonometry as well.

Exercise

1. Check to see that our new definition allows us to write

$$\begin{aligned} h_b &= c \sin \alpha &= a \sin \gamma , \\ h_c &= b \sin \alpha &= a \sin \beta , \end{aligned}$$

as a cyclic substitution would produce.

5 The Law of Sines

In a triangle, we have two expressions for h_c:

$$h_c = a \sin \beta = b \sin \alpha .$$

We obtain an interesting relationship if we divide the last two equal quantities by the product $\sin \alpha \sin \beta$:

$$\frac{a \sin \beta}{\sin \alpha \sin \beta} = \frac{b \sin \alpha}{\sin \alpha \sin \beta} ,$$

or

$$\frac{a}{\sin \alpha} = \frac{b}{\sin \beta}.$$

And we can get corresponding proportions from the other pairs of sides by making the same cyclic substitution as before. We obtain

$$\frac{a}{\sin \alpha} = \frac{b}{\sin \beta} = \frac{c}{\sin \gamma}.$$

This is a very important relationship among the sides and angles of a triangle. It is known as the *Law of Sines*.

This formula has many interesting connections. For example, you may have learned in geometry that if two sides of a triangle are unequal, then the greater side lies opposite the greater angle: If $\beta < \alpha$ then $AC < BC$.

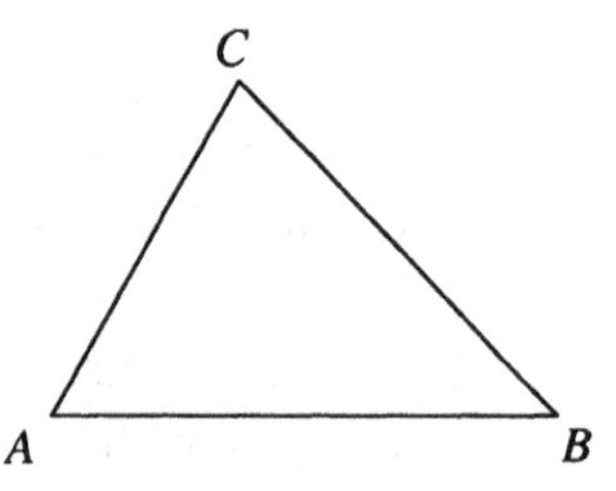

But it is *not* true that if angle β is double the angle α, then side BC is double the side AC. This is shown clearly in the figure below, with a 30-60-90 triangle. As we know, there is a side double the smallest, but it's not the one opposite the 60 degree angle. It's the hypotenuse.

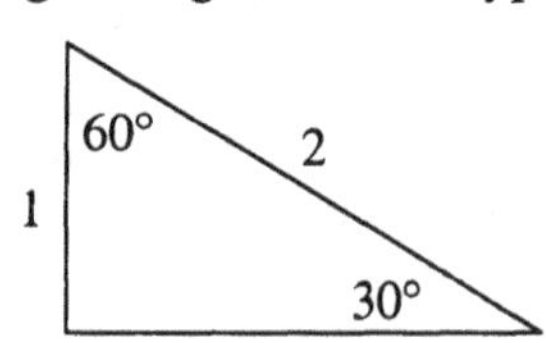

The Law of Sines generalizes correctly the fact that the greater side lies opposite the greater angle, because it tells us that the ratio of two sides of a triangle is the ratio of the sines of the opposite angles. And, as we have seen, the sines of two angles are not in the same ratio as their degree-measures.

The Law of Sines can help us in another way too, which we mentioned at the start of this chapter. We know from geometry that two triangles are congruent if two pairs of corresponding angles are equal, and a pair of corresponding sides are equal (in many textbooks, this is called ASA or SAA, depending on whether the side is between or outside of the two angles). Another way to say this is to assert that the measure of two angles and one side determines the triangle. Geometry shows us one way to get the third angle (using the fact that the three angles of a triangle sum to 180°). But geometric methods do not let us compute the lengths of the other two sides. The law of sines allows us to do this.

Exercises

1. Verify that the cyclic substitutions give the equalities shown above.
2. Check that the Law of Sines holds in a 30-60-90 triangle.
3. Use the Law of Sines in the triangles below to determine the lengths of the missing sides. (Use your calculator for the computations.)

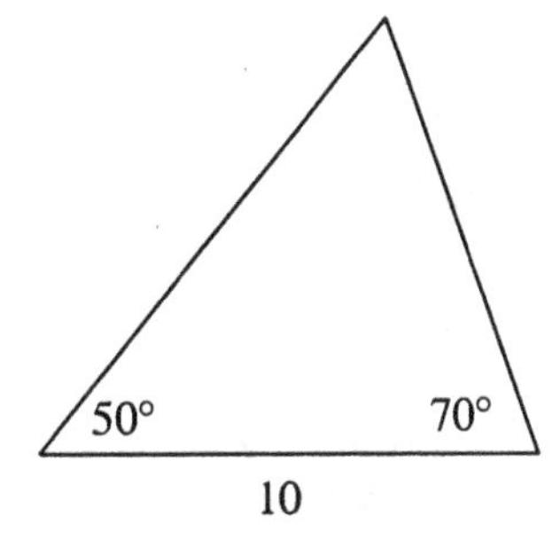

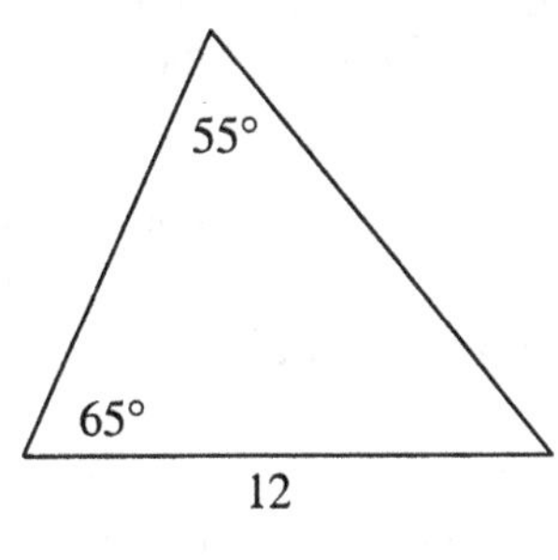

4. We have defined the sine of an obtuse angle as equal to the sine of its supplement. With this definition, show that the law of sines is true for an obtuse triangle.
5. Use the Law of Sines in the triangles below to determine the lengths of the missing sides.

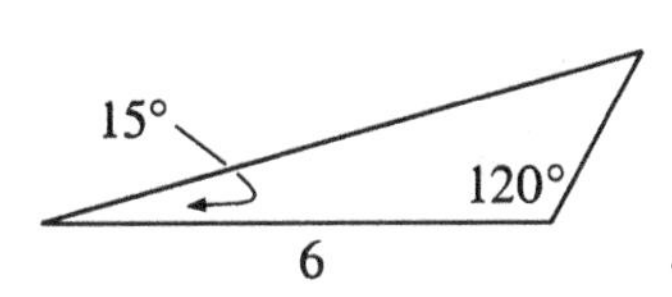

6. Use the Law of Sines to find the two missing angles in the triangle below:

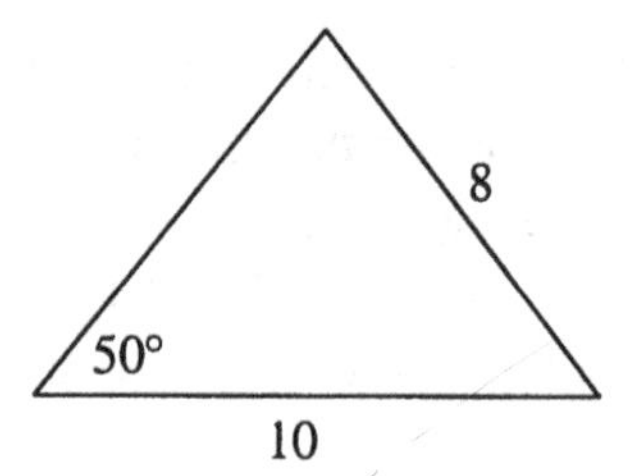

7. Recall from geometry that SSA does not guarantee congruence. That is, if two triangles match in two sides and an angle not included between these two sides, then the triangles may not be congruent. Look back at Problem 6. Is the triangle determined uniquely? How many possible values are there for the degree measurements of the remaining angles?

8. Suppose triangle ABC is inscribed in a circle of radius R. Prove the *extended Law of Sines*:

$$\frac{a}{\sin A} = \frac{b}{\sin B} = \frac{c}{\sin C} = 2R\,.$$

6 The circumradius

We can learn more about the Law of Sines another way if we give a geometric interpretation of the ratio $a/\sin\alpha$ in any triangle ABC.

We construct the circle circumscribing the triangle:[2]

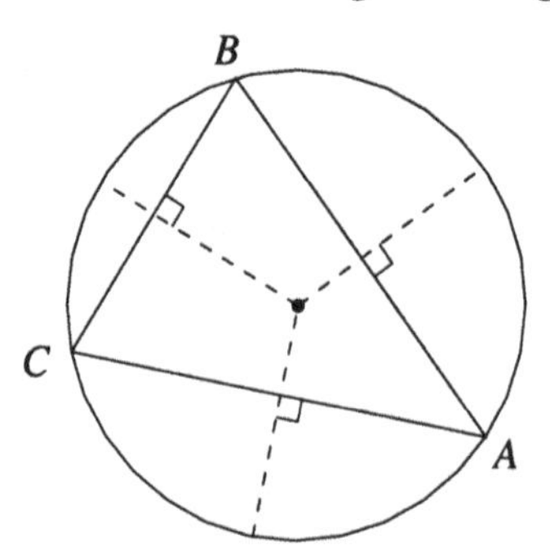

Suppose the radius of this circle (the *circumradius* of the triangle) is R. We know from the result on page 57 that

$$BC = a = 2R\sin\alpha\,.$$

[2]Recall that the perpendicular bisectors of the three sides of a triangle coincide at a point equidistant from all three vertices. This point is the center of the triangle's circumscribed circle.

So the ratio $a/\sin\alpha$ is simply equal to $2R$.

Exercises

1. Find the circumradius of a triangle in which a 30° angle lies opposite a side of length 10 units. Note that this information does not determine the triangle.

2. Find the circumradius of a 30-60-90 triangle with hypotenuse 8. Do you really need the result of this section to find this circumradius?

7 Area of a triangle

Our altitude formulas have given us one interesting result: the Law of Sines. We now show how they lead to a new formula for the area of a triangle. But in fact, the formula we present is not really new. It is just the usual formula from geometry, written in trigonometric form.

If S denotes the area of a triangle, we know that

$$S = \frac{1}{2}ah_a\,.$$

But $h_a = b\sin\gamma$, so we can write

$$S = \frac{1}{2}ab\sin\gamma\,.$$

This is our "new" formula. As with our other formulas, we can use "cyclic substitutions" (see page 69) to get two more formulas:

$$S = \frac{1}{2}bc\sin\alpha$$

$$S = \frac{1}{2}ca\sin\beta\,.$$

Exercises

1. Find the area of a triangle in which two sides of length 8 and 11 include an angle of 40° between them.

2. Find the areas of the triangles shown:

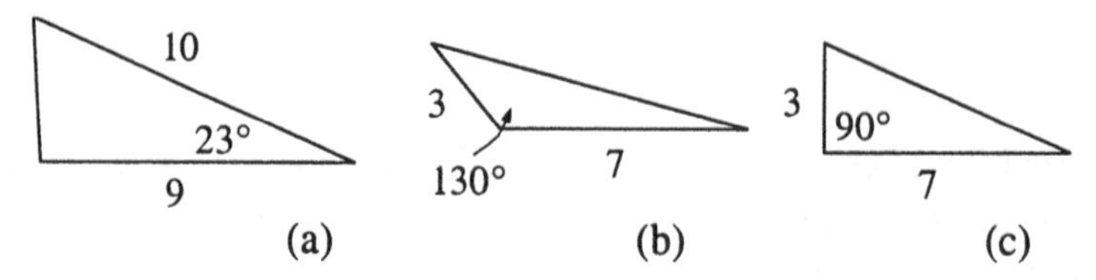

Can you use our new formula for part (c)? Is it necessary to use this formula?

3. The area of triangle ABC is 40. If side AB is 6 and angle A is 40 degrees, find the length of side AC.

4. In triangle PQR, side $PQ = 5$, and side $PR = 6$. If the area of the triangle is 9, find the degree-measure of angle P.

   Hint: There are two possible answers. Can you find them both?

5. Two sides of a triangle are a and b. What is the largest area the triangle can have? What is the shape of the triangle with largest area?

   Answer: The largest area is $ab/2$, achieved when the angle between the two sides is a right angle.

   Challenge: There is another right triangle with sides a and b. Find this triangle and its area.

6. The length of a leg of an isosceles triangle is x. Express in terms of x the largest possible area the triangle can have.

7. Show that the area of a parallelogram is $ab \sin C$, where a and b are two adjacent sides and C is one of the angles. Does it matter which angle we use?

8. We start with any quadrilateral whose diagonals are contained inside the figure. Show that the area of the quadrilateral is equal to half the product of the diagonals times the sine of the angle between the diagonals. Should we take the acute angle formed by the diagonals, or the obtuse angle?

9. Show that we can use the same formula to get the area of a quadrilateral whose diagonals (when extended) intersect outside the figure.

10. In the figure, $AD = 4$, $AE = 6$, $AB = 8$, $AC = 10$. Find the ratio of the area of triangle ADE to that of triangle ABC.

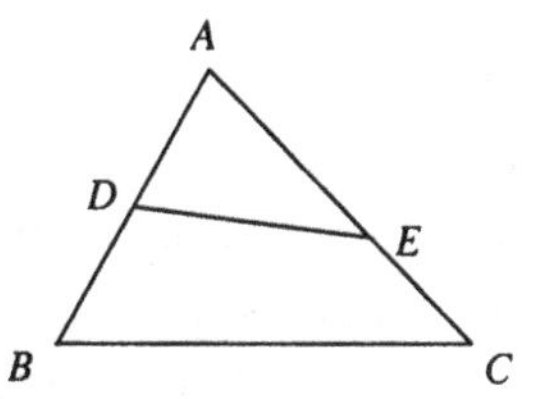

11. In quadrilateral $PQRS$, diagonals PR and QS intersect at point T. The sum of the areas of triangles PQT and RST is equal to the sum of the areas of triangles PST and QRT. Show that T is the midpoint of (at least) one of the quadrilateral's diagonals.

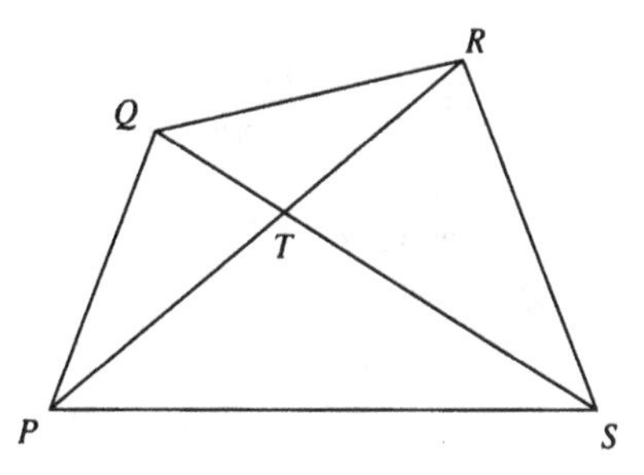

Solution. The sines of angles PTQ, QTR, RTS, STP are all equal. If this sine is s, and using absolute value for area, we have $|PQT| + |RST| = (1/2)PT \times QT \times s + (1/2)ST \times RT \times s = |QRT| + |PTS| = (1/2)QT \times RT \times s + (1/2)PT \times ST \times s$, so $PT \times QT + RT \times ST = QT \times RT + PT \times ST$, or $PT \times QT + RT \times ST - QT \times RT - PT \times ST = 0$, or $(PT - RT)(QT - ST) = 0$. But this means that one of the factors must be zero, so that T is the midpoint of at least one of the diagonals. □

12. In quadrilateral $ABCD$, diagonals AC and BD meet at point P. Again using absolute value for area, show that $|APB| \times |CPD| = |BPC| \times |DPA|$. Is this true if the intersection point of the diagonals is outside the quadrilateral?

13. In acute triangle ABC, show that $c = a \cos B + b \cos A$.

    Hint: Draw the altitude to side c. How must we change this result if angle A or angle B is obtuse?

8 Two remarks

Remark 1: Note that these formulas express the area of a triangle in terms of two sides and an included angle. We knew from geometry that these three pieces of information determine the triangle (and therefore its area), but we need trigonometry to actually compute the area. We will see in the next section how trigonometry allows us to compute other elements of a triangle determined by two sides and their include angle.

Remark 2: We now have three ways to think of $\sin\alpha$ geometrically:

1. In a right triangle with an acute angle α, $\sin\alpha$ is the ratio of the leg opposite α to the hypotenuse:
$$\sin\alpha = \frac{\text{opposite leg}}{\text{hypotenuse}}.$$
2. In any triangle, $\sin\alpha$ is the ratio of the side opposite α to the diameter of the circumscribed circle:
$$\sin\alpha = \frac{a}{2R}.$$
3. In any triangle, $\sin\alpha$ is the ratio of twice the area to the product of the two sides which include α:
$$\sin\alpha = \frac{2S}{bc}.$$

We can use whichever fits the situation we are working in. Indeed, it turns out that any of these could actually function as the definition of $\sin\alpha$.

9 Law of cosines

The law of cosines is a very old theorem. It appears in Euclid's *Elements*, the very first textbook of geometry, although Euclid does not use the term cosine. It is a generalization of the Pythagorean theorem.

In triangle ABC, if angle B is an acute angle, then

$$b^2 = a^2 + c^2 - 2ac\cos B.$$

In fact, this is not difficult to prove:

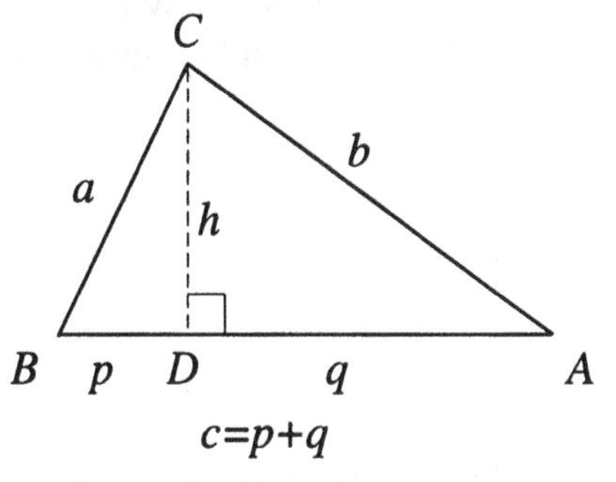

From right triangle BDC, we have $p = a\cos B$. Using the Pythagorean theorem twice, in triangles ACD and BCD, we have $b^2 = h^2 + q^2 = a^2 - p^2 + (c-p)^2 = a^2 + c^2 - 2pc = a^2 + c^2 - 2ca\cos B$, which is what we wanted to prove.

The Pythagorean theorem says that the square of a side of a triangle opposite a right angle is equal to the sum of the squares of the other two sides.

One of the ways in which the law of cosines generalizes the Pythagorean theorem is by showing that the square of a side of a triangle opposite an acute angle is less than the sum of the squares of the other two sides.

What if we take the side of a triangle opposite an obtuse angle?

Exercise Show that if b is a side opposite an obtuse angle of a triangle, then $b^2 = a^2 + c^2 + 2ac\cos B'$, where B' is the measure of the supplement of obtuse angle B.

(A hint is contained in the diagram below.)

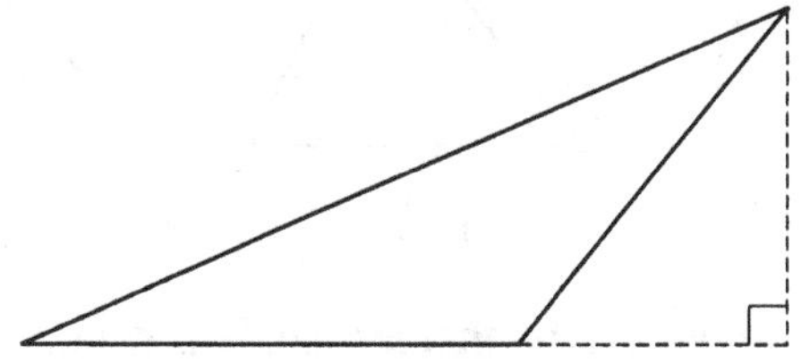

On the basis of this result, we make a second daring definition (to follow our daring definition of the sine of an obtuse angle):

Definition The cosine of an obtuse angle is the cosine of its supplement, multiplied by -1.

So we have three results: the Pythagorean theorem for a right angle, and the two new results for an acute and an obtuse angle. Just as with the sine function, we can make all these results into a single formula, the so-called *Law of Cosines*: In triangle ABC,

$$b^2 = a^2 + c^2 - 2ac\cos B\,.$$

Exercises

1. Check to see that this is correct, whether angle B is acute, right, or obtuse.

2. In each of the triangles below, use the Law of Cosines to express the square of the indicated side in terms of the other two sides and their included angle:

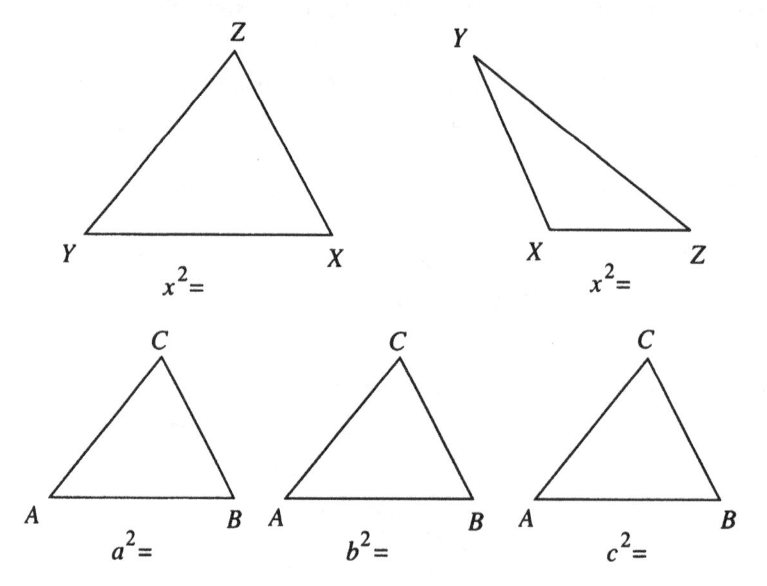

3. We know, from geometry, that a triangle is determined by SAS (the lengths of two sides and the angle between them). Explain how the Law of Cosines allows us to calculate the missing parts of a triangle, if we are given SAS.

4. Find the side or angle marked x in each diagram below:

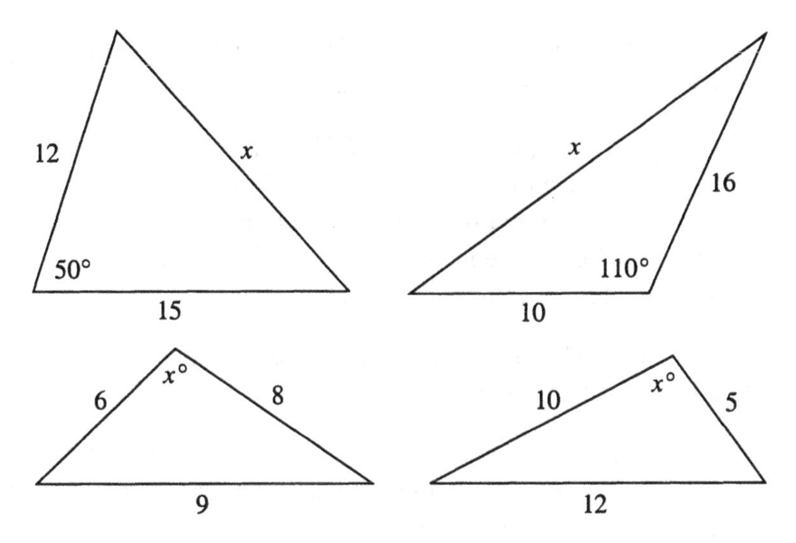

5. In triangle ABC, $AB = 10$, $AC = 7$, and $BC = 6$. Find the measures of each angle of the triangle.

6. Peter's teacher gave the following problem:

   A parallelogram has sides 3 and 12. Find the sum of the squares of its diagonals.

   But Peter had trouble even drawing the diagram. He knew that opposite sides of a parallelogram are equal, so he knew where to put the numbers 3 and 12. But then he didn't know what kind of parallelogram to draw. He drew a rectangle (which, he knew, is a kind of parallelogram). Then he drew a parallelogram with a 30° angle, and another parallelogram with a 60° angle. But he didn't know which one to use to do the computation.

   Can you help Peter out?

7. Show that the sum of the squares of the sides of any parallelogram is equal to the sum of the squares of the diagonals.

8. If M is the midpoint of side BC in triangle ABC, then AM is called a median of triangle ABC. Show that for median AM, $4AM^2 = 2AB^2 + 2AC^2 - BC^2$.

   Hint: The diagram for this problem is "half" of the diagram for Exercise 7 above.

9. Show that the sum of the squares of the three medians of a triangle is 3/4 the sum of the squares of its sides.

10. The diagonals of quadrilateral $ABCD$ intersect inside the figure. Show that the sum of the squares of the sides of the quadrilateral is equal to the sum of the squares of its diagonals, plus four times the length of the line segment connecting the midpoints of the diagonals (notice that this generalizes problem 6).

11. In triangle ABC, angle C measures 60 degrees, $a = 1$ and $b = 4$. Find the length of side c.

12. In triangle ABC, angle C measures 60 degrees. Show that $c^2 = a^2 + b^2 - ab$. What is the corresponding result for triangles in which angle C measures 120 degrees?

13. Three riders on horseback start from a point X and travel along three different roads. The roads form three 120° angles at point X. The first rider travels at a speed of 60 MPH, the second at a speed of 40 MPH, and the third at a speed of 20 MPH. How far apart is each pair of riders after 1 hour? After 2 hours?

Appendix – Three big ideas and how we can use them

I. Invariants: Motions in the plane

We often talk about the congruence of triangles. Two triangles are congruent if one can be moved so that it fits exactly on the other. So we can say that two congruent triangles are exactly the same, except for their position.

The two triangles below cannot be considered congruent if we confine our motions to the plane. To move one of them onto the other, we must flip it around (reflect it in a line) before we can make it fit. These triangles are *mirror images* of each other.

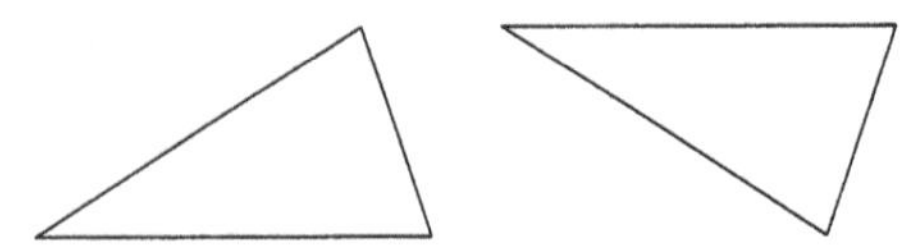

Exercises

1. Most triangles cannot be placed on their mirror images without reflecting them in a line. However, there are certain special triangles that can be placed onto their mirror images without using reflections at all. Draw one such triangle.

2. Describe the set of all triangles that can be placed onto their mirror images without reflection in a line.

3. Draw some quadrilaterals that can be placed onto their mirror images without reflection in a line.

I.1 Triangle invariants

A triangle invariant is a quantity associated with a triangle that is unaffected by its position. Thus the value of a triangle invariant for any two congruent triangles will be equal. Some examples of triangle invariants are the lengths of the sides, the measures of the angles, and the area. While these seven invariants are basic, there are many others (such as the lengths of the altitudes, or the radius of the circumscribed circle).

When we work with the relationships among triangle invariants, we are connecting the geometry of the triangle with algebra and trigonometry. A mathematician would say that algebra and trigonometry are the *analytical tools* of geometry.

I.2 The sine and triangle invariants

We can give an alternative definition of the sine of an angle in terms of triangle invariants. Indeed, we have already seen how.

We have seen (Chapter 3, section 7) that for any triangle,

$$S = \frac{1}{2}ab\sin\gamma$$

where γ is the angle included between two sides of lengths a and b. So if we start with any angle γ, draw a triangle (not necessarily a right triangle!) including it, and denote by a and b the measures of the sides surrounding γ, then we can define $\sin\gamma$ as equal to $2S/ab$, where S is the area of the triangle.

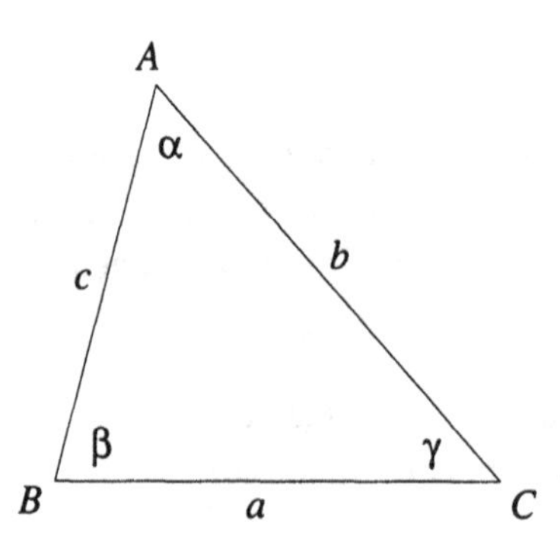

Whether you think of this statement as a consequence of our original development, or as a definition of the sine of an angle, is your choice. In either case, we have the following:

$$\sin\alpha = \frac{2S}{bc}, \quad \sin\beta = \frac{2S}{ac}, \quad \sin\gamma = \frac{2S}{ab}.$$

Exercises

1. Note that if α is a right angle in the diagram above, then the area of $\triangle ABC$ is $bc/2$. Show that in this case, the formulas for $\sin\beta$ and $\sin\gamma$ given above are just what were given in Chapter 1.

2. Using the law of cosines, show that in any triangle ABC of area S,

$$c^2 = a^2 + b^2 - 4S\cot\gamma.$$

II. Symmetry

Let us look once more at the formula which expresses the area of a triangle s in terms of two sides and the sine of the included angle. We already know that there are three of these formulas:

$$S = \frac{1}{2}ab\sin\gamma, \quad S = \frac{1}{2}bc\sin\alpha, \quad S = \frac{1}{2}ca\sin\beta.$$

Each of them is obtained from the others by *cyclic substitutions* of a, b and c, and α, β and γ, respectively.

In general, when we have a formula for a triangle, we can expect this sort of symmetry. No one of the sides and angles plays a special role with respect to the others, so if in the formula we perform a cyclic substitution of them, we should get a valid formula as well.

We can also look at this process in reverse. We can consider the three equal quantities:

$$ab \sin \gamma = bc \sin \alpha = ca \sin \beta .$$

When we have three symmetric expressions that are equal, sometimes we can find a geometric reason why they are all equal. In this case, they are all equal to twice the area of the triangle.

In applying cyclic substitutions, we must be sure that each variable in our formula can represent *any* side or angle in a triangle. For instance, the Pythagorean theorem says that if a and b are the legs of a right triangle, and c the hypotenuse, then $a^2 + b^2 = c^2$. We cannot subsitute a for b, b for c, and c for a, because c cannot be any side of the triangle: it must be the hypotenuse. (However we can substitute a for b and b for a.)

Exercises

1. The law of cosines says that

$$c^2 = a^2 + b^2 - 2ab \cos \gamma .$$

Using cyclic substitutions, write down two more formulas like this one.

2. The law of sines says that

$$\frac{a}{\sin \alpha} = \frac{b}{\sin \beta} = \frac{c}{\sin \gamma} .$$

Can you find a geometric reason why these three quantities are equal? **Hint:** See page 74.

III. The sine and its dimension

Physicists often deal with dimensions as well as numbers, since the numbers they use are often the result of some measurement. For example, sides of triangles are measured in units of length (such as centimeters), while their areas are measured in units of length squared (such as square centimeters). We can borrow this idea from the physicist, by noting that in an algebraic or trigonometric identity, both sides should have the same dimension. For example, in the Pythagorean theorem, the dimension of both sides is length squared, the same dimension as areas. And in fact, our proof of this theorem interpreted it as a statement about areas (as does the proof in Euclid's *Elements*).

What is the dimension of $\sin\alpha$? As we originally defined it, $\sin\alpha$ is the ratio of two lengths, so in fact its dimension is 0. (This is another way of saying that the unit of length used to measure the sides of a triangle does not affect the value of the sines of its angles.)

Let us check that the dimensions are correct in our new formulas. We have written $\sin\alpha = 2S/bc$. Now S has dimension length squared, and the product bc has the same dimension (length times length), so the dimensions cancel out, and $\sin\alpha$ has dimension 0. This agrees with our previous result.

Exercise

1. Check that the dimensions of each side are the same in the following formulas:

   a) $\dfrac{a}{\sin\alpha} = \dfrac{b}{\sin\beta}$.

   b) $\dfrac{a}{b} = \dfrac{\sin\alpha}{\sin\beta}$.

   c) $S = \frac{1}{2}ab\sin\gamma$.

   d) $c^2 = a^2 + b^2 - 2ab\cos\gamma$.

IV. Hero's formula

We know that any two triangles with the same three side lengths are congruent. This means that they will give the same value for any triangle invariant, such as the area. That is, the lengths of the sides of a triangle *determine* its area.

There is a wonderful formula, credited to Hero (or Heron) of Alexandria, which expresses the area of a triangle in terms of the lengths of its sides. If these lengths are a, b and c, and if $s = (a+b+c)/2$, we have that

$$S = \sqrt{s(s-a)(s-b)(s-c)}\,.$$

Let us prove this formula.

We know that $\sin\gamma = 2S/ab$, so that

$$\sin^2\gamma = \frac{4S^2}{a^2b^2}\,.$$

From the law of cosines, we have

$$c^2 = a^2 + b^2 - 2ab\cos\gamma\,,$$

or

$$\cos\gamma = \frac{a^2 + b^2 - c^2}{2ab}\,.$$

Hence,

$$\cos^2\gamma = \frac{(a^2 + b^2 - c^2)^2}{4a^2b^2}\,.$$

Finally, we remember that $\sin^2\gamma + \cos^2\gamma = 1$, and now we have all that we need. If we substitute the results above for $\sin^2\gamma + \cos^2\gamma$, we will have a relationship that includes only S, a, b, and c, just what we want.

Indeed,

$$\sin^2\gamma + \cos^2\gamma = \frac{4S^2}{a^2b^2} + \frac{(a^2 + b^2 - c^2)^2}{4a^2b^2} = 1\,,$$

or

$$16S^2 + (a^2 + b^2 - c^2)^2 = 4a^2b^2\,,$$

or

$$16S^2 = 4a^2b^2 - (a^2 + b^2 - c^2)^2\,.$$

This is the relationship we need, but it doesn't look very "nice." In particular, it doesn't look symmetric in a, b and c.

But in fact it is. We can show this by factoring the right-hand side as the difference of two squares:

$$\begin{aligned}16S^2 &= 4a^2b^2 - (a^2 + b^2 - c^2)^2\\ &= \big(2ab + (a^2 + b^2 - c^2)\big)\big(2ab - (a^2 + b^2 - c^2)\big)\,,\end{aligned}$$

and so

$$\begin{aligned}16S^2 &= \big(2ab + (a^2 + b^2 - c^2)\big)\big(2ab - (a^2 + b^2 - c^2)\big)\\ &= (a^2 + 2ab + b^2 - c^2)(-a^2 + 2ab - b^2 + c^2)\,.\end{aligned}$$

Each of the factors above on the right is again the difference of two squares:

$$16S^2 = \big((a+b)^2 - c^2\big)\big(c^2 - (a-b)^2\big)\,,$$

so we can factor once more:

$$\begin{aligned} 16S^2 &= ((a+b)+c)((a+b)-c)(c+(a-b))(c-(a-b)) \\ &= (a+b+c)(a+b-c)(a-b+c)(-a+b+c), \end{aligned}$$

and we now see the beautiful symmetry of the expression. We can write this relationship as

$$S^2 = \frac{(a+b+c)(a+b-c)(c-a+b)(c+a-b)}{16},$$

or

$$S = \frac{\sqrt{(a+b+c)(a+b-c)(c-a+b)(c+a-b)}}{4}.$$

This formula is nice, but it can be made even nicer if we set $s = (a+b+c)/2$. Then we have:

$$\begin{aligned} a+b-c &= 2s-2c, \\ c-a+b &= 2s-2a, \\ c+a-b &= 2s-2b. \end{aligned}$$

Substituting these results into the formula above, we then obtain

$$S = \frac{\sqrt{(2s)2(s-a)2(s-b)2(s-c)}}{4} = \sqrt{s(s-a)(s-b)(s-c)}.$$

Exercises

1. Show that Hero's formula gives the correct value for the area of a triangle with sides 3, 4 and 5.

2. Show that Hero's formula gives the correct value for the area of a triangle with sides 5, 12 and 13.

3. Using Hero's formula, show that the area of an equilateral triangle with side of length l is given by $l^2\sqrt{3}/4$.

4. Show that the formula $S = \frac{1}{2}ab\sin\gamma$ also leads to the formula in Problem 3 above.

5. Use Hero's formula to solve Problems 9 and 10 on page 10.

V. A physicist's interpretation

Hero's formula may seem strange. In most area formulas, you multiply just two quantities together, but here we multiply four quantities together. We make up for it by taking a square root, but this is also unusual for an area formula.

To help us make a bit more sense of this formula, we can imagine Richard Feynman, a Nobel Laureate in Physics, who was very skilled at explaining subtle ideas simply. He might have explained Hero's formula in the following way:

In high school I had a very good course in geometry, and I remember studying Hero's formula, which relates the lengths of the side of a triangle to its area. But I've forgotten its exact form. Let's see what I can recall. I know it had a square root in it.[3] *Now the dimension of the area is length squared, so under the square root we must have a polynomial of degree four.*[4] *We can get such a polynomial by multiplying together four factors, each of degree* 1.

What could these factors be? Well, if $a + b = c$, *then our* triangle *is actually a line segment (as the triangle inequality tell us), which has area* 0. *So when* $a + b - c = 0$, *the whole polynomial is zero. This means that* $a+b-c$ *is a factor of the polynomial. Similarly,* $a-b+c$ *must be a factor, and so must be* $-a + b + c$.

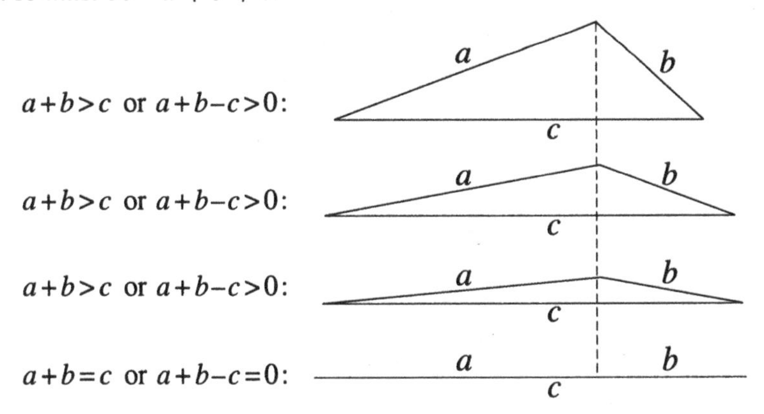

[3] In fact, Feynman's mathematician friends could explain why there must be a square root in the formula. The explanation involves attaching a sign to the triangle's area, depending on the orientation of the triangle.

[4] For a polynomial of several variables, the degree of each term is the sum of the exponents of all the variables that appear in it, and the degree of the polynomial is the highest of the degrees of its terms.

So we have three of the four factors of the polynomial under the square root. What can the fourth factor be? It must be linear in a, b and c, and it must be symmetric in these three variables.[5] *This means that the fourth factor must be of the form $k(a+b+c)$, for some constant k.*

So we must have

$$S = C\sqrt{(a+b+c)(a+b-c)(a-b+c)(-a+b+c)}\,,$$

for some constant C. We can determine this constant by examining one particular triangle, and I remember that a triangle with sides 3, 4 *and* 5 *is a right triangle. The area of this triangle is* 6, *and the expression*

$$C\sqrt{(a+b+c)(a+b-c)(a-b+c)(-a+b+c)}$$

has the value $C\sqrt{(12)(2)(4)(6)} = C \times 24$ for this particular triangle. Therefore, $C = 6/24 = 1/4$. I also remember that we can clean this up algebraically by introducing $s = (a+b+c)/2$, but I will leave this to my friends the mathematicians. And now that I've had fun figuring out what the formula must be, I also leave to them the actual proof. They are good at that.

[5]What Feynman would mean here is that if we interchange any two of the variables a, b and c, the value of the polynomial would be unaffected.

Chapter 4

Angles and Rotations

1 Measuring rotations

In previous chapters we explored the meaning of expressions such as $\sin 30^\circ$, $\cos 45^\circ$ and $\tan 60^\circ$. In this chapter and the next we show how we can use expressions such as $\sin 180^\circ$, $\tan 300^\circ$ or even $\sin 1000^\circ$.

But what might 1000° measure? Certainly it is not the measure of the angle of a triangle. These can only be between 0° and 180° (acute, right or obtuse). Nor can it be the measure of an angle (or an arc) in a circle. These can only be between 0° and 360°.

If you have ever owned toy electric trains, you may have set up the tracks in a circle, and run the trains around the circle. The diagram below shows a circular track. If a train starts at point A, travels around the circle, and arrives back at point A, we say that it has made one full rotation around the circle.

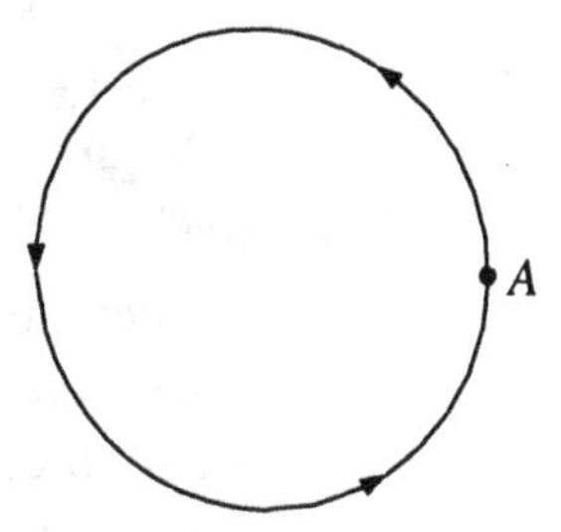

Since we divide a circle into 360 degrees, it is natural to say that the train has rotated around the circle by 360 degrees.

Now suppose the train continues past point A, and travels around the circle again. Then we can say that it has rotated through more than 360 degrees. If it travels around the circle twice, returning to point A, we say that it has rotated $360 + 360 = 720$ degrees.

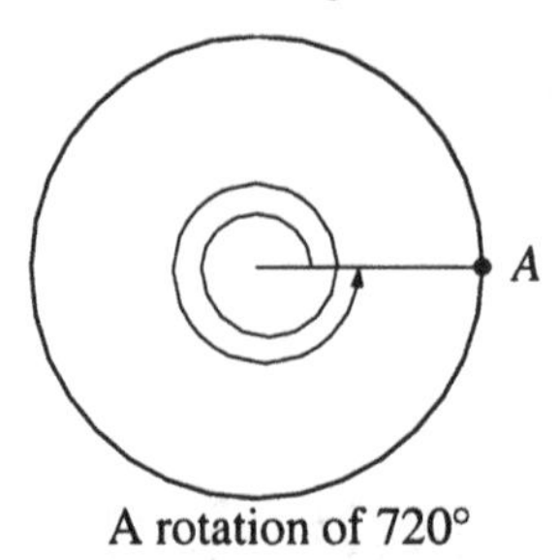

A rotation of 720°

And if it travels a bit further around the circle, along an arc measuring 280°, we say that it has rotated $720° + 280° = 1000°$:

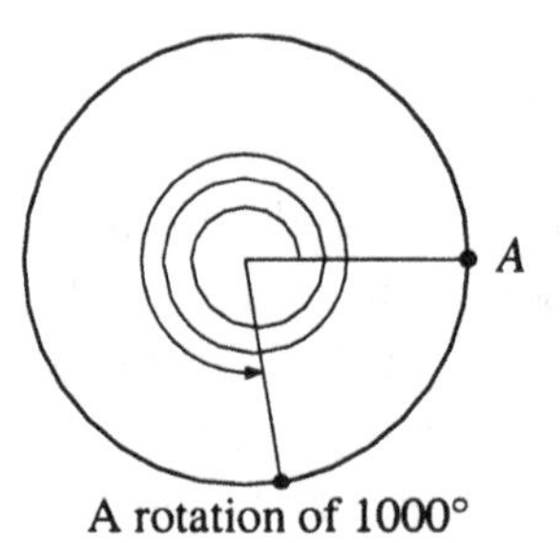

A rotation of 1000°

Here is another example. Look at the hour hand of a clock. In 12 hours it has made a full rotation, or rotated by 360°.

But this time the rotation is clockwise (by definition!), while our train was rotating counterclockwise. In a plane, there are two different directions of rotation, and it turns out to be important to distinguish between them. Mathematicians call a counterclockwise rotation *positive* and a clockwise rotation *negative*. So we say that in 12 hours, the hour hand of a clock performs a rotation of $-360°$.

Exercise

1. Draw diagrams showing the following rotations:

a) $160°$	b) $190°$	c) $400°$
d) $600°$	e) $1200°$	f) $-70°$
g) $-400°$	h) $360°$	i) $-270°$

2 Rotation and angles

Picture a circle of radius 1, with its center at the origin of a system of coordinates:

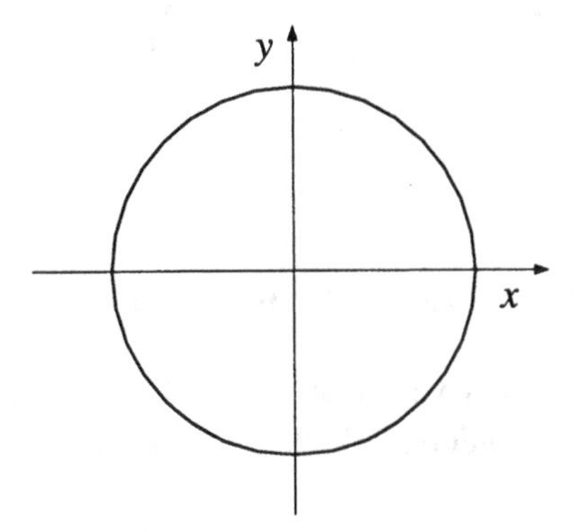

We take an acute angle with one leg along the x-axis. The other leg will end up someplace in the first quadrant. If the measure of this acute angle is, say, $40°$, then we can get from point Q to point P by rotating through $40°$.

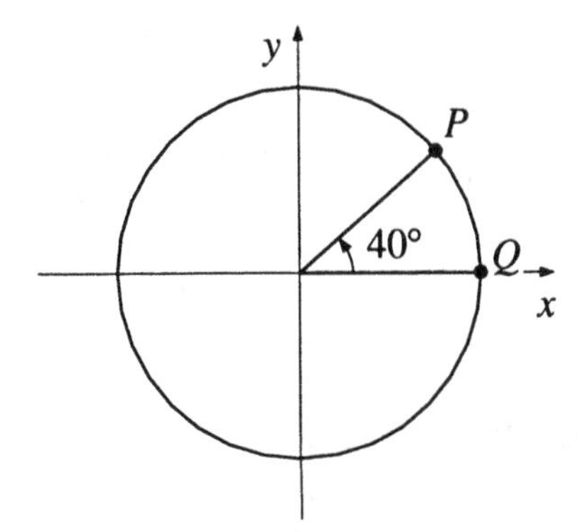

So we can associate angles with rotations. Even if the rotation exceeds 180°, we sometimes talk about the "angle" instead of the "rotation." The figure below gives some examples.

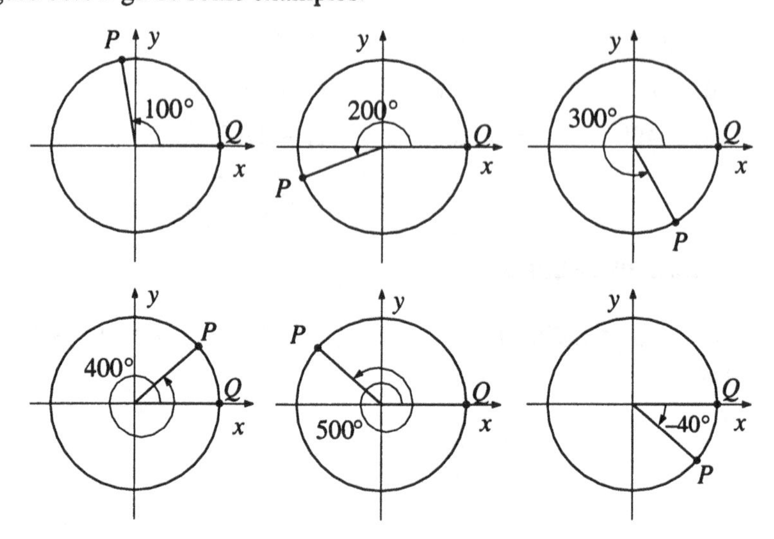

3 Trigonometric functions for all angles

Let us look again at what we mean by $\sin 40°$. We will do so in such a way that it will help us understand what is meant by $\sin 300°$, $\cos 1100°$, or $\tan(-240°)$.

We draw a circle of radius r centered at the origin of coordinates. To find $\sin 40°$, we mark the point P in the first quadrant such that $\angle POR = 40°$, and drop the perpendicular PR to the x-axis:

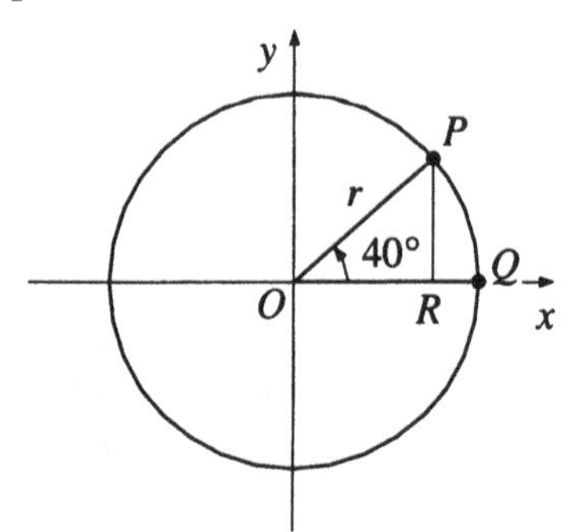

From right triangle POR, we see that

$$\sin 40° = \frac{PR}{OP} = \frac{PR}{r}.$$

Similarly, we can write

$$\cos 40^\circ = \frac{OR}{r}.$$

But if the coordinates of P are (x, y), we see that $y = PR$ and $x = OR$. So we can write

$$\sin 40^\circ = \frac{y}{r}, \quad \cos 40^\circ = \frac{x}{r}.$$

So far we have said nothing new.

Or have we?

We can use this observation to extend our definitions of sine and cosine to our new angles, which measure rotations. Suppose a point P starts at position $(r, 0)$, and rotates through an angle α.

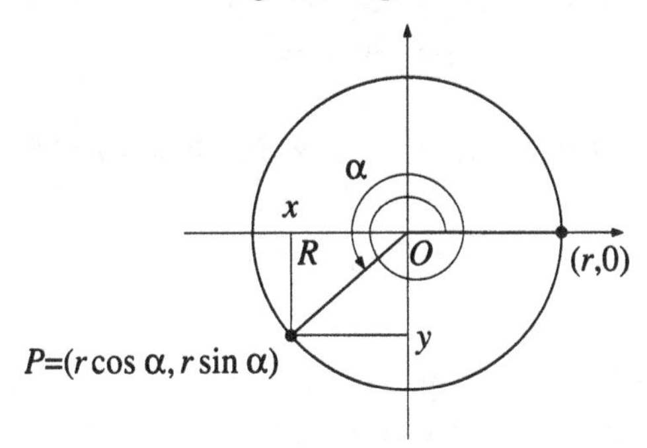

If P has coordinates (x, y), we define $\cos\alpha$ and $\sin\alpha$ by writing

$$\cos\alpha = \frac{x}{r},$$
$$\sin\alpha = \frac{y}{r}.$$

Note that these new definitions give the same values as the old definitions when α is an acute angle.

Example 29 Find the numerical values of $\sin 130^\circ$ and $\cos 130^\circ$.

Note that by Chapter 3, page 71, we already know that $\sin 130^\circ = \sin(180^\circ - 130^\circ) = \sin 50^\circ$, so in fact, this quantity had been defined already. But let us see if our new definition gives the same result.

Solution. The diagram below shows $\angle QOP = 130^\circ$:

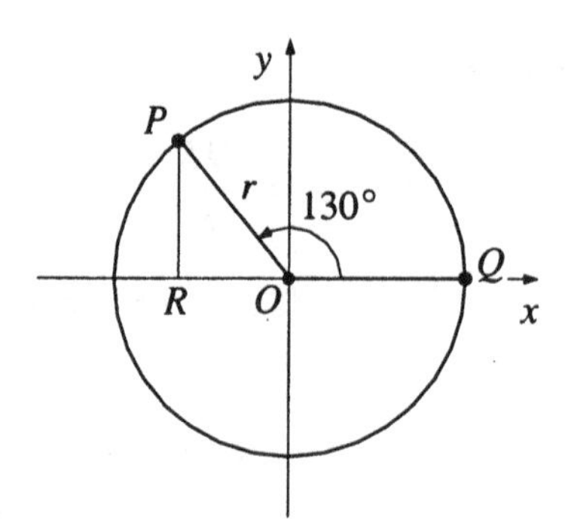

The circle in the diagram has radius r, and the point P has rotated through an angle of 130° from point Q. If the coordinates of P are (x, y), our new definition tells us that

$$\cos 130^\circ = \frac{x}{r},$$
$$\sin 130^\circ = \frac{y}{r}.$$

Now we look at right triangle OPR, in which $\angle POR = 50^\circ$, and note that

$$\frac{y}{r} = \frac{PR}{OP} = \sin 50^\circ .$$

So this is the value of $\sin 130^\circ$.

Triangle OPR will also give us the numerical value of $\cos 130^\circ$, but we must be careful. Since the x-coordinate of point P is negative, we must write

$$\cos 130^\circ = \frac{x}{r} = -\frac{OR}{OP} = -\cos 50^\circ .$$

Thus, $\cos 130^\circ = -\cos(180^\circ - 130^\circ) = -\cos 50^\circ$. This value agrees with the one given by the definition on page 79. □

It is important to note that the result above does not depend on the length of OP. We can choose a circle of any radius and draw the corresponding diagram for a 130° angle. Triangle OPR will always have the same angles, and the computation will be the same.

Example 30 What are the values of $\cos 210^\circ$ and $\sin 210^\circ$?

Solution. Since the radius of the circle will not matter, we are free to choose, for example, a circle of radius 1. Then our new definitions lead to the diagram below:

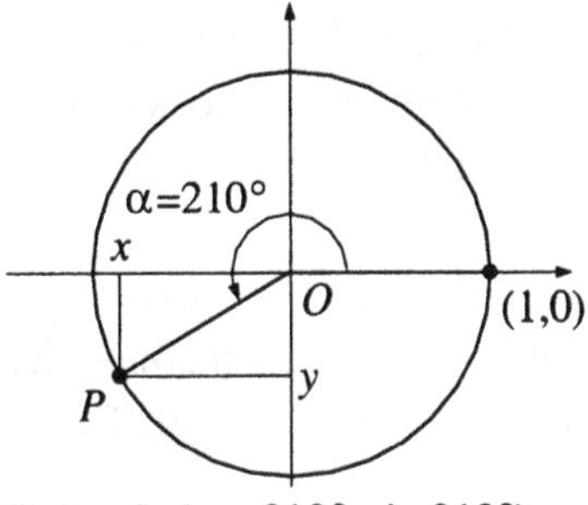

$P=(x,y)=(\cos 210°, \sin 210°)$

The geometry of a 30-60-90 triangle shows that the coordinates of point P are $(-\frac{\sqrt{3}}{2}, -\frac{1}{2})$. Then

$$\cos 210° = -\frac{\sqrt{3}}{2} \qquad \text{and} \qquad \sin 210° = -\frac{1}{2}.$$

Notice that both the sine and cosine of 210° are negative numbers. □

Example 31 Find the values of $\cos 360°$ and $\sin 360°$.

Solution. We choose a circle of radius 1. For a rotation of 360°, the coordinates of point P are $(1, 0)$. Therefore, $\cos 360° = 1$ and $\sin 360° = 0$. □

Now that we have definitions for sine and cosine of any angle, we can make definitions for the other trigonometric functions of these angles.

For any angle α,

$$\begin{aligned} \tan\alpha &= \frac{\sin\alpha}{\cos\alpha}, \\ \cot\alpha &= \frac{\cos\alpha}{\sin\alpha}, \\ \sec\alpha &= \frac{1}{\cos\alpha}, \\ \csc\alpha &= \frac{1}{\sin\alpha}. \end{aligned}$$

Example 32 Find the numerical value of $\tan 210°$.

Solution. From the results of Example 30, we have that

$$\tan 210° = \frac{\sin 210°}{\cos 210°} = \frac{-1/2}{-\sqrt{3}/2} = \frac{1}{\sqrt{3}}.$$

Note that this value is positive. □

Our new definitions of sine and cosine give values for any angle α. But this is not quite true for our new definitions of tangent, cotangent, secant and cosecant, because they involve division. We must be sure that we are not dividing by 0.

Indeed, we will not define $\tan\alpha$ if $\cos\alpha = 0$. Expressions such as $\tan 90°$, $\tan 270°$, and $\tan(-90°)$ must remain undefined.

For similar reasons, we cannot define $\cot 0°$, or $\csc 180°$.

Exercises

1. Find the numerical value of the following expressions. Do this without using your calculator, then check your answers with your calculator.

   a) $\sin 390°$ b) $\cos 3720°$ c) $\tan 1845°$

   d) $\sin 315°$ e) $\cot 420°$ f) $\tan(-30°)$

2. Find the numerical value of each expression below, or indicate if the given expression is undefined.

   a) $\tan 360°$ b) $\sin 180°$ c) $\cos 180°$

   d) $\cot 90°$ e) $\cot 360°$ f) $\tan(-270°)$

4 Calculations with angles of rotations

Let us look back at our original picture of an angle in a circle:

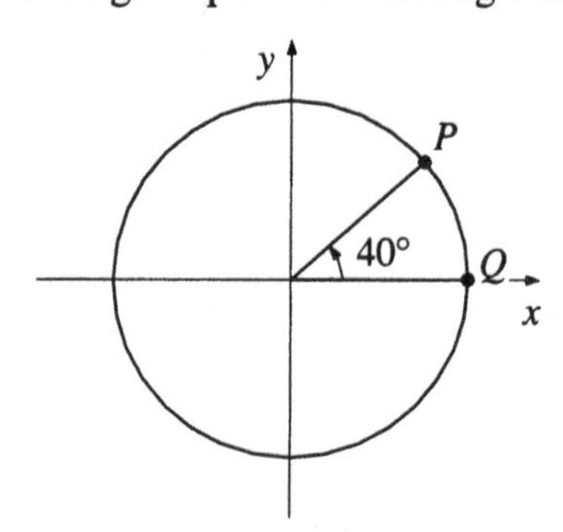

Originally, we thought of this as an angle of 40°. But a diagram of a 400° angle would look exactly the same, as would a diagram for 760° or −320°.

The diagram will look the same for any two angles which differ by a full rotation. Therefore, $\sin\alpha = \sin(\alpha + 360°)$.

Similarly, $\cos\alpha = \cos(\alpha + 360°)$ for any angle α.

These observations allow us to find the sine, cosine, tangent, or cotangent of very large angles easily.

Example 33 What is $\cos 1140°$?

Solution. If we divide 1140 by 360, the quotient is 3 and the remainder is 60, that is, $1140 = 3 \cdot 360 + 60$. So $\cos 1140° = \cos(3 \cdot 360 + 60°) = \cos 60° = 1/2$. □

Example 34 Is the sine of 100,000° positive or negative?

Solution. If we divide 100,000 by 360, we get 277, with a remainder of 280. So the sine of 100,000° is the same as $\sin 280°$. Since the position of point P is in the fourth quadrant, its y-coordinate is a negative number. The sine of 100,000° is therefore negative. □

You can check the logic of these solutions using your calculator, which already "knows" if the sine of an angle is positive or negative. That is, the people who designed it did exercises like yours before they built the calculator.

But it is also important to be able to "predict" certain values of the trigonometric functions, or at least tell whether their values will be positive or negative. It's not difficult to see that if point P ends up in quadrant I, all functions of the angles are positive. If point P lies in quadrant II, the sine and cosecant are positive, and all other functions are negative, and so on.

Exercise Check to see that the diagrams below give the correct signs for the functions of angles in each quadrant.

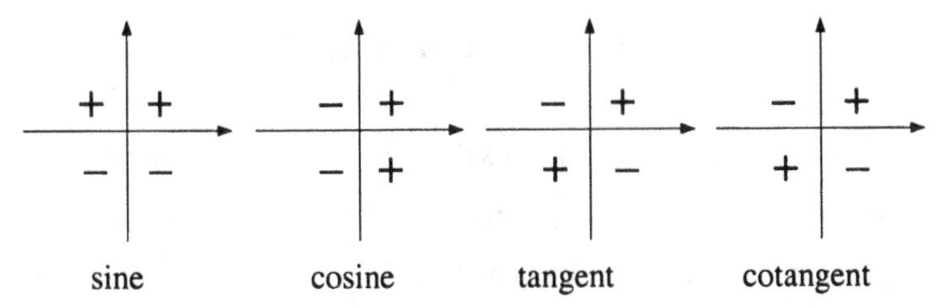

We can even, sometimes, predict a bit more about the values of the trigonometric functions. If you look at each of the diagrams below, you may see that $\sin\theta$ is equal, in absolute value, to the sine of the acute angle made by one side of the angle and the x-axis (the angle marked γ in each diagram below):

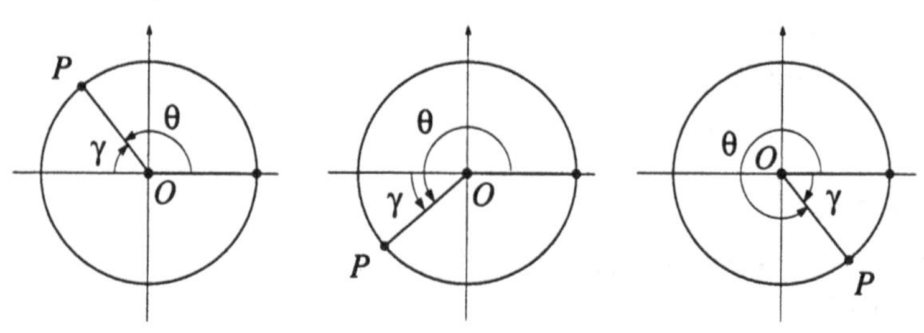

Example 35 Find $\sin 300^\circ$.

Solution. Point P, having rotated through 300°, will end up in quadrant IV. So $\sin 300^\circ$ is negative. Furthermore, the angle made by one side and the x-axis is 60°. Hence $\sin 300^\circ = -\sin 60^\circ = -\sqrt{3}/2$. □

Exercises

1. In what quadrant will the point P lie after a rotation of 400°? 3600°? 1845°? -30°? -359°?

2. Fill in the table below (you won't need a calculator). What is the relationship between $\sin\alpha$ and $\sin(-\alpha)$?

$\sin 30^\circ$		$\sin(-30^\circ)$	
$\sin 135^\circ$		$\sin(-135^\circ)$	
$\sin 210^\circ$		$\sin(-210^\circ)$	
$\sin 300^\circ$		$\sin(-300^\circ)$	
$\sin 390^\circ$		$\sin(-390^\circ)$	
$\sin 480^\circ$		$\sin(-480^\circ)$	

3. Solve the following equations for α, where $0 < \alpha < 360^\circ$:

a) $\sin\alpha = 0$ b) $\cos\alpha = 0$ c) $\sin\alpha = 1$

d) $\cos\alpha = 1$ e) $\sin\alpha = -1$ f) $\cos\alpha = \frac{1}{2}$

g) $\sin\alpha = -\frac{1}{2}$ h) $\sin^2\alpha = \frac{1}{2}$ i) $\cos^2\alpha = -\frac{3}{4}$

4. a) If $\sin\alpha = 5/13$, in what quadrant can α lie? What are the possible values of $\cos\alpha$?

b) If $\sin\alpha = -5/13$, in what quadrant can α lie? What are the possible values of $\cos\alpha$?

5. We have seen (Chapter 2, page 50) that if a and b are non-negative numbers such that $a^2 + b^2 = 1$, then there exists an angle θ such that $\sin\theta = a$ and $\cos\theta = b$. Show that this is true, even if a or b is negative.

5 Odd and even functions

Consider the result of Exercise 2 on page 100. If you have filled in the table correctly, you will note that, for the angles given there, $\sin\alpha$ and $\sin(-\alpha)$ have opposite signs. This relationship holds in general:

$$\sin(-\alpha) = -\sin\alpha \quad \text{for any angle } \alpha\,.$$

Similarly, we find the following:

$$\begin{aligned}\tan(-\alpha) &= -\tan\alpha\\ \cot(-\alpha) &= -\cot\alpha\,.\end{aligned}$$

However, the cosine function is different. We have

$$\cos(-\alpha) = \cos\alpha\,.$$

In general, we can distinguish two type of functions.

A function is called **even** *if, for every* x, $f(-x) = f(x)$.

A function is called **odd** *if, for every* x, $f(-x) = -f(x)$.

So, for example, the functions

$$f(x) = \cos x\,, \quad f(x) = x^2 + 3\,, \quad \text{and } f(x) = \frac{1}{x^6}$$

are all even, while the functions

$$f(x) = \tan x\,, \quad f(x) = x^3 + 4x\,, \quad \text{and } f(x) = \frac{1}{x^7}$$

are all odd. The following functions are neither even nor odd:

$$f(x) = x^3 + x^2\,, \quad f(x) = \sin x + \cos x\,.$$

In summary:

$\cos x$ *is an even function, while* $\sin x$, $\tan x$, *and* $\cot x$ *are odd functions.*

This may be the reason that some mathematicians prefer to work with the cosine function, rather than the sine.

Exercises

1. Which of the following functions are even? Which are odd? Which are neither?

   a) $f(x) = x^6 - x^2 + 7$
   b) $f(x) = x^3 - \sin x$
   c) $f(x) = \frac{1}{x+1}$
   d) $f(x) = \sec x$
   e) $f(x) = \csc x$
   f) $f(x) = 2 \sin x \cos x$
   g) $f(x) = \sin^2 x$
   h) $f(x) = \cos^2 x$
   i) $f(x) = \sin^2 x + \cos^2 x$

2. If $f(x)$ is any function, show that

   $$g(x) = \tfrac{1}{2}(f(x) + f(-x))$$

   is an even function, and that

   $$h(x) = \tfrac{1}{2}(f(x) - f(-x))$$

   is an odd function. Use these results to show that every function can be written as the sum of an even and an odd function.

3. Express the following functions as the sum of an even and odd function:

   a) $f(x) = \sin x + \cos x$
   b) $f(x) = x^3 + x^2 + x + 1$
   c) $f(x) = 2^x$
   d) $f(x) = \dfrac{1 - \sin x}{1 + \sin x}$
   e) $f(x) = \dfrac{1}{x + 2}$

Chapter 5

Radian Measure

1 Radian measure for angles and rotations

So far, our unit of measurement for angles and rotations is the degree. We measure an angle in degrees using a *protractor*:

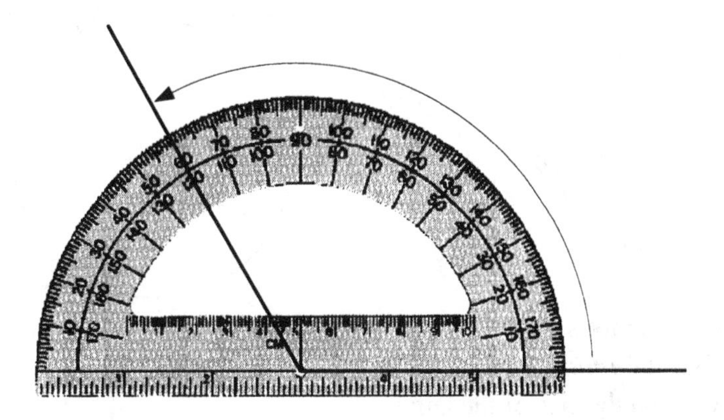

Why are there 360 degrees in a full rotation? The answer to this lies in history, not in mathematics. It turns out that there is a more convenient way to measure angles and rotations called *radian* measure. Mathematicians and scientists find it natural to use radian measure to express relationships. Unlike degree measure, radian measure does not depend on an arbitrary unit.

To measure an angle in radians, we place its vertex at the center of any circle, and think about the length of arc AB, as measured in inches, centimeters, or some other unit of length:

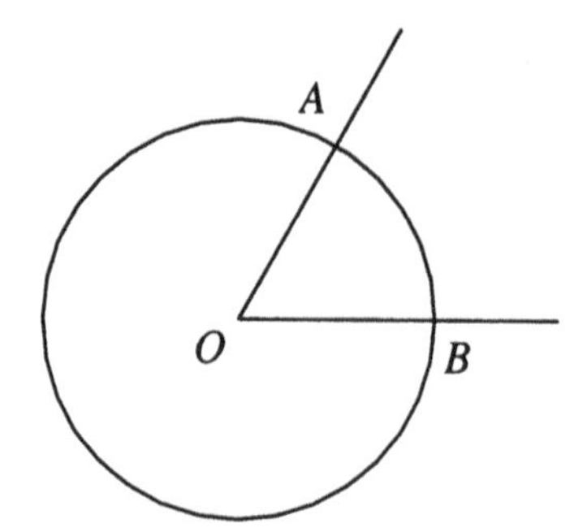

The length of this arc depends on the size of angle AOB:

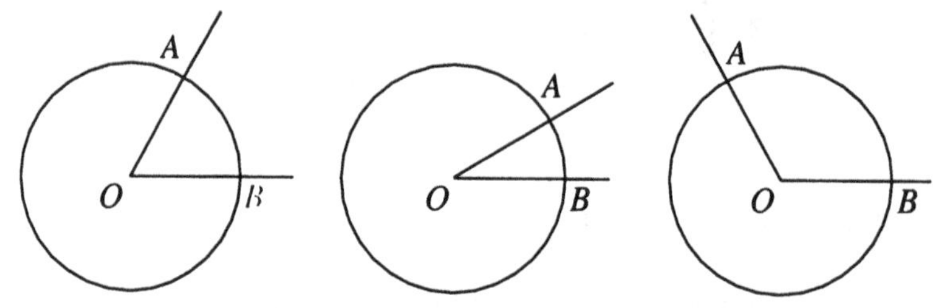

But is also depends on the size of the radius of the circle:

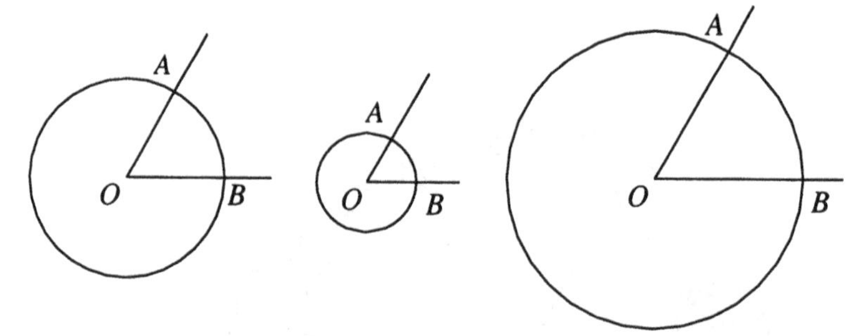

So we cannot simply take the length of this arc as the measure of the angle. But the *ratio* of the length of the arc to the radius of the circle depends only on the size of the angle:

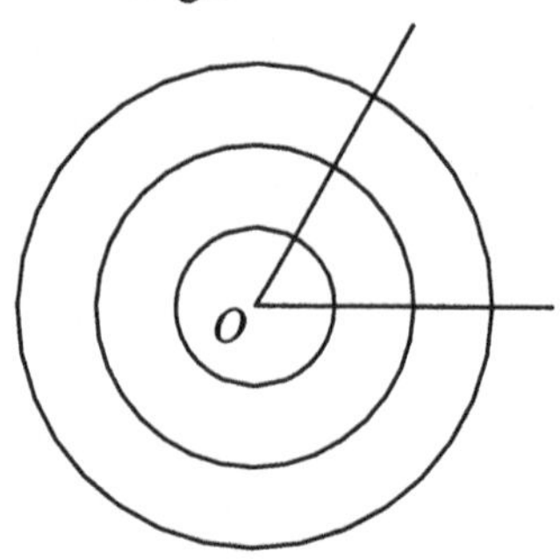

Definition The radian measure of an angle is the ratio of the arc it cuts off to the radius of any circle whose center is the vertex of the angle.[1]

This definition reminds us of the definition of the sine of an angle, which is also a ratio, and which does not depend on the particular right triangle that the angle belongs to.

Example 36 What is the radian measure of an angle of 60°?

Answer. We place the vertex of our 60° angle at the center of a circle of radius r, and examine the arc it cuts off. Since $60/360 = 1/6$, this arc is 1/6 of the circumference of the circle. So its length is $(1/6)(2\pi r) = \pi r/3$ units.[2] By our definition, the radian measurement of our 60° is the ratio

$$\frac{\pi r/3}{r} = \frac{\pi}{3}.$$

Numerically, this is approximately 1.0471976, or a little more than 1 radian. □

Example 37 What is the radian measure of an angle of 360°?

Answer. $2\pi r/r = 2\pi$ radians. □

Example 38 What is the degree measure of an angle of 1 radian?

Answer. An angle of 2π radians is 360 degrees (see Example 37). So an angle of 1 radian is $360/2\pi = 180/\pi$ degrees. □

In Example 38, we have used the very important fact that radian measure is *proportional* to degree measure. In fact, it is not hard to see that

---

[1] There are two simple tests that this measurement passes. First, the bigger the angle, the bigger its radian measure. Second, if we place two angles next to each other (see figure),

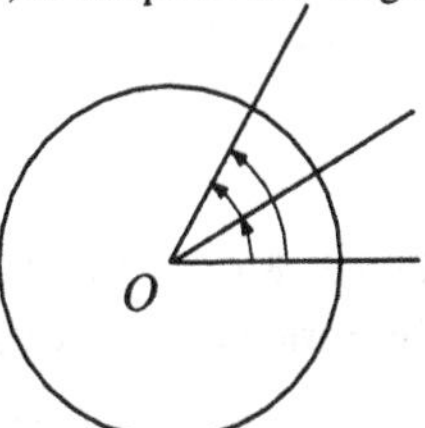

the measure of the larger angle they form together is the sum of the measures of the two original angles.

[2] Recall that if r is the radius of a circle, the length of its circumference is given by the formula $2\pi r$.

the ratio of the radian measure of an angle to its degree measure is always $\pi/180$:

$$\boxed{\frac{\text{radians}}{\text{degrees}} = \frac{\pi}{180}}$$

In general, if we are measuring an angle in radians, we do not use any special symbol like the "degree" sign.

Example 39 Find $\sin(\pi/6)$.

Solution. We first express the angle in degrees. If D is the required degree measure, we have

$$\frac{\text{radians}}{\text{degrees}} = \frac{\pi/6}{D} = \frac{\pi}{180},$$

which leads to $D = 30°$, and we know that $\sin 30° = 1/2$. □

Example 40 In a circle of radius 1, what is the length of the arc cut off by a central angle of 2 radians?

Solution 1 (the long and hard way). We saw above that the degree-measure of this angle is about 114°. So the arc cut off by this angle is approximately

$$\frac{114}{360} \times 2\pi \approx 1.989675347274$$

units long. □

Solution 2 (the neat and easy way). In a unit circle (whose radius is 1 unit), the radian measure of a central angle is just the length of the arc it cuts off. This tells us that the required arc is exactly two units long (and gives us an idea of the error we made in using the approximate degree measure in Solution 1). □

Example 41 A central angle in a circle of radius 2 units cuts off an arc 5 units long. What is the radian measure of this angle?

Solution. By definition, this radian measure is $5/2$. □

Exercises

1. What is the radian measure of an angle of 180°? 90°?
2. What is the degree measure of an angle of 2 radians?
3. What is the radian measure of $1/4$ of a full rotation?
4. What is the radian measure of a rotation through an angle of 45°?
5. Fill in the following table:

Degree measure	Radian Measure
90	
180	
270	
360	
	$\pi/2$
	π
	$3\pi/2$
	2π

6. Fill in the following tables:

Degree measure	Radian measure
0	
30	
72	
120	
135	
	$\pi/6$
	$\pi/5$
	$\pi/4$
	$\pi/3$
	$2\pi/3$
	$7\pi/10$

Degree measure	Radian measure
198	
210	
216	
225	
240	
	$11\pi/10$
	$10\pi/9$
	$7\pi/6$
	$6\pi/5$
	$5\pi/4$
	$4\pi/3$

7. What is the radian measure of an angle of 1 degree?

8. Using your calculator, find the sine of an angle of (a) 1 radian; (b) 1 degree.

9. Without using your calculator, fill in the following table:

α (in radian)	$\sin\alpha$	$\cos\alpha$
$\pi/6$		
$\pi/3$		
$\pi/2$		
$2\pi/3$		
$7\pi/6$		
$5\pi/4$		
$3\pi/2$		
$11\pi/6$		

10. In a circle of radius 1, what is the length of an arc cut off by a central angle of 2 radians? Of 3 radians? Of π radians?

11. In a circle of radius 3, what is the length of an arc cut off by a central angle of 2 radians? Of 3 radians? Of π radians?

12. If $\sin\pi/9 = \cos\alpha$ and α is acute, what is the radian measure of α?

13. If α is an angle between 0 and $\pi/2$ (in radian measure), which is bigger: $\sin\alpha$ or $\cos(\pi/2 - \alpha)$?

14. Let us take an angle whose radian measure is 1. Using the picture below, prove that its degree-measure is less than 60°. (In fact, an angle of radian measure 1 is approximately 57 degrees.)

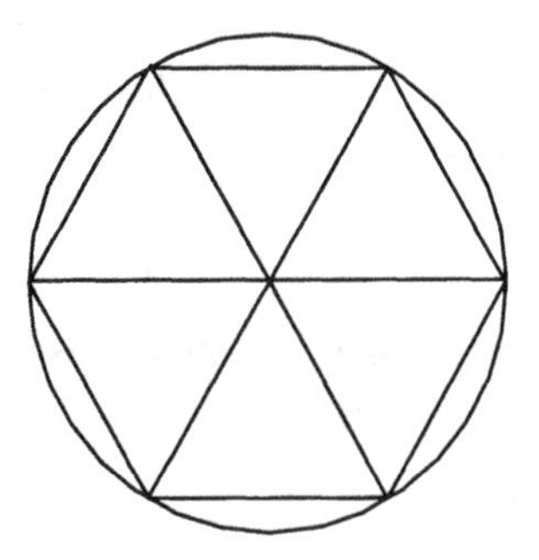

2 Radian measure and distance

Imagine a wheel whose radius is 1 foot. Let this wheel roll, without slipping, along a straight road:[3]

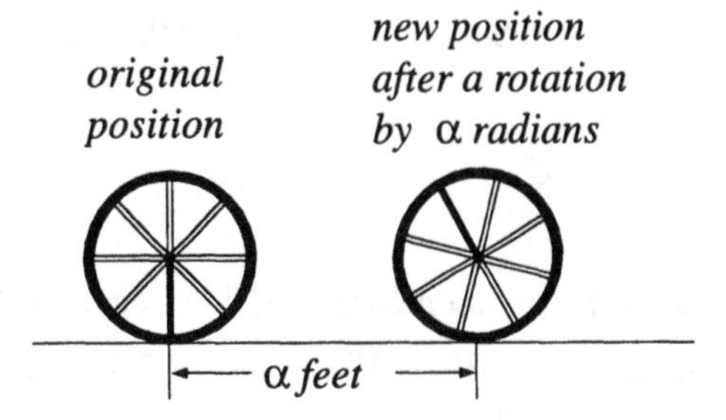

Since the wheel does not slip as it rolls, the distance it rolls, in feet, is just the length of the arc that the angle α cuts off.

Example 42 How far will a wheel of radius 1 foot travel after 1 rotation?

Solution. Because it rotates without slipping, the wheel will travel exactly the length of the circumference of the circle. But if the radius of the circle is r, then the circumference is $2\pi r$. Since $r = 1$, the answer is just 2π, or approximately 6.28 feet. □

[3]Sometimes a car wheel slips as it rolls. This is called a *skid*, and it happens when there is not enough friction between the wheel and the road (for example when the road is icy or wet). A car wheel can also turn without rolling: sometimes, a car stuck in deep snow will *spin its wheels*. We assume that neither of these things is happening to our wheel.

Example 43 A wheel of radius 1 foot rotates through $1/2$ a rotation. How far will it travel?

Answer. It will travel a distance of π, or about 3.14 feet. □

Example 44 How far does a wheel of radius 1 foot travel along a line, if it rotates through an angle of 2 radians?

Answer. Two feet.

Example 45 Through how many radians does a circle of radius 1 foot rotate, if it travels 5 feet down a road?

Answer. Five radians.

Example 46 How much does a wheel with radius 1 foot rotate if it travels 1000 feet along a road? Give the answer in radians and also in degrees.

Solution. In radians, this is easy: it has rotated through 1000 radians.

In degrees, the answer is more difficult to find. Each full rotation covers 2π feet. So in traveling 1000 feet, our wheel has rotated through $1000/2\pi \approx 159.155$ rotations. Since each rotation is $360°$, the degree measure of a rotation of 1000 radians is

$$159.155 \times 360 \approx 57296° .$$ □

Example 47 What is the radian measure of an angle of $1000°$?

Solution. A rotation of $1000°$ is $1000/360 \approx 2.77$ rotations, and each rotation is 2π radians. So $1000°$ is

$$2.77 \times 2\pi \approx 17.405$$

in radian measure. □

Example 48 Is $\sin 500$ (in radian measure) a positive or a negative number?

Solution. Since $500/2\pi \approx 79.577$, 500 radians is 79 full rotations, plus approximately 0.57 of one more rotation. Since $0.5 < 0.57 < 0.75$, this is between 1/2 and 3/4 of one rotation. So a rotation of 500 radians will end up in the third quadrant, and its sine is a negative number. □

Exercises

1. Through how many radians does a circle of radius 1 foot rotate, if it travels 5 feet down a road?

2. Through how many degrees does a circle of radius 1 foot rotate, if it travels 5 feet down a road?

3. How far does a circle of radius 1 foot travel, if it turns through an angle of 4 radians?

4. How far does a circle of radius 1 foot travel, if it turns through an angle of $120°$?

5. In a circle of radius 1, what is the length of an arc cut off by an angle with radian measure 1/2? $\pi/2$? α?

6. What is the radian measure of an angle of 720 degrees? 1440 degrees? 3600 degrees? 15120 degrees? What is the degree measure of an angle whose radian measure is 12π? 15π? 100π?

7. In a circle of radius 3, what is the length of an arc cut off by an angle with radian measure 1/2? $\pi/2$? α?

8. In a circle of radius 3, how long is the arc cut off by an angle with radian measure 1.5?

9. In a circle of radius 5, how long is the arc cut off by an angle of 80 degrees?

10. In a circle of radius 2, what is the radian measure of a central angle whose arc has length 3 units?

11. In a circle of radius 6, what is the degree-measure of a central angle whose arc has length 2 units?

12. A circle of radius 7 units rolls along a straight line. If it covers a distance of 20 units, what is the radian measure of the rotation it has made?

13. A circle of radius 8 rolls along a straight line, through an angle of 150 degrees. How far does it roll?

14. Through what angle does the hour hand of a watch rotate in one hour? Give your answer in radians.

15. Through what angle does the minute hand of a watch rotate in one hour? The second hand?

16. In answering problem 14, Joe Blugg gave the following solution: One hour on the face of a watch is 1/12 of the circle, so it is $2\pi/12 = \pi/6$ radians. In degrees, the answer is $360/12 = 30$ degrees. But Joe is *not* correct, either in degrees or in radians. Find and correct his mistake. Did you make the same error here and in similar problems about a watch?

    Hint: Do the hands of a clock rotate? In which direction?

17. Let us look at a pocket watch whose hour hand is exactly one inch long. Suppose the tip of this hour hand travels a distance of 1000 inches as it goes around. How long does this trip take?

18. Suppose the length of the hour hand of Big Ben is exactly one yard long. How long it will it take Big Ben's hour hand to turn through 1000 radians?

19. A wheel with radius 1 meter is rolling along a straight line. One of its spokes is painted red. At the starting position this spoke is vertical, with its endpoint towards the ground. How many radians does the wheel turn before the spoke is again in this position? How many radians does the wheel turn before the spoke is vertical, with its endpoint towards the sky?

20. A wheel whose radius is 1 meter rolls along a straight path. The path is marked out in 3-meter lengths, with red dots three meters apart. The wheel has a wet spot of blue paint on one point. When it starts rolling, that point is touching the ground. As the wheel rolls, it leaves a blue mark every time the initial point touches the ground again.

    a) How far apart are the blue marks?

    b) Through what angle has the wheel rolled between the time it makes a blue mark and the time it makes the next blue mark?

    c) Will a blue mark ever coincide with a red mark?

d) When the blue marks do not coincide with the red marks, how close do they come to the red marks? (If you know how to program a computer or calculator, you may need to write a program to answer this question.)

e) Now suppose each interval between the red dots is divided into four equal sub-intervals, say by pink dots in between. A blue mark is created as the wheel completes its 100th rotation. Between what two dots does this blue mark occur?

3 Interlude: How to explain radian measure to your younger brother or sister

When you drive with Mom or Dad in the car, did you ever notice the odometer? That's the little row of numbers in front of the steering wheel. It measures the distance covered by the car, in miles.

But how does it know this? The odometer cannot read the road signs, telling us how far we've come. It must get the information from the car's wheels. But the car's wheels can only tell the odometer how much they have turned. The more the car's wheels turn, the more distance we cover. The odometer knows how to convert rotations to miles. In geometry, we learn that a circle of radius r has a circumference of $2\pi r$. This, and the radius of the wheel in feet, is all the odometer really needs to know.

Suppose the wheel tells the odometer that it has rotated 50 times. Then the odometer knows that the wheel has traveled $50 \times 2\pi r$ feet, where r is the radius of the car's wheels in feet (it must convert this number to miles). And if the wheel has rolled only 1/4 of the way around, the odometer reports a distance of $(1/4) \times 2\pi r$ feet, again with a conversion to miles.

But suppose you want to know how much wear the tires have had. Then we must read the odometer, and figure out how many rotations the tires have made from the distance they traveled. So if the odometer says that the car has traveled 200 feet (we have to convert from miles, again), then 200 is $2\pi r$ times the amount of rotation the wheels have made. So the wheels have made $200/2\pi r$ rotations.

And this is what we call the radian measure of this rotation.

4 Radian measure and calculators

Most calculators, and all scientific calculators, know about radian measure. You can switch your calculator between "degree mode" and "radian mode" (and sometimes there are still other ways to measure angles). But each calculator does this in a different way. It is important that you know how to tell which mode your calculator is working in, and also how to switch from one mode to another.

Exercises

1. A student asked his calculator for the sin of 1. The answer was 0.8414709848079. Was the calculator in radian mode or in degree mode?

2. For small angles, $\sin x$ is approximately equal to x, when x is given in radian measure. Use your calculator to find out how big the difference is between x and $\sin x$ for angles of radian measure 0.2, 0.15, 0.05.

   In each case, which is bigger, x or $\sin x$?

3. A better approximation to $\sin x$ (when measured in radians) is given by $x - x^3/6$. Find the difference between this value and the actual value of $\sin x$ for the three angles above.

4. In the old schools of artillery, the officers would use a version of the approximation $\sin x \approx x$ However, they had to measure x in degrees, so they used $\sin x = x/60$. What is the error in this approximation, if $x = 10°$?

5. a) Without your calculator, make a guess for the value of $\sin 0.1$ (in radian measure). Then use a calculator to check your guess.

   b) Now perform the same experiment for $\sin 0.1$ (in degree measure).

6. a) Find the sine of an angle whose degree measure is 1000.

   b) Find the sine of an angle whose radian measure is 1000.

7. a) Find sin (sin 1000), where radian measure is used for the angle.

   b) Find $\sin 3.14$, where radian measure is used for the angle.

8. Without looking up this number on your calculator, prove that $\cos 1.5707$ is less than 0.0001.

   Hint: Do you recognize the number 1.5707?

5 An important graph

Let us summarize our knowledge of the sine function by drawing its graph.

The integer multiples of π will give us a convenient scale for the x-axis, since the values of $\sin x$ at these points are easy to calculate. For the y-axis, we need only values from -1 to 1, since $\sin x$ can only take on these values.

We can draw the graph by looking at a unit circle (drawn on the right below), and recording the height of a point which makes an angle α with the x-axis. Here is what it looks like for a typical acute angle α.

As α varies from 0 to $\pi/2$, the graph of $y = \sin x$ increases.

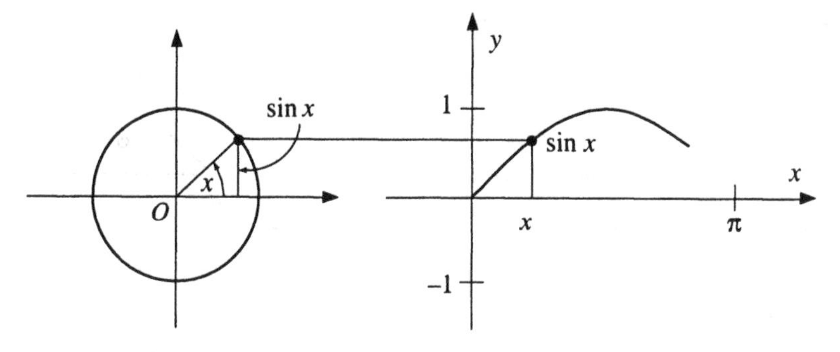

Here is a typical scene from the second quadrant:

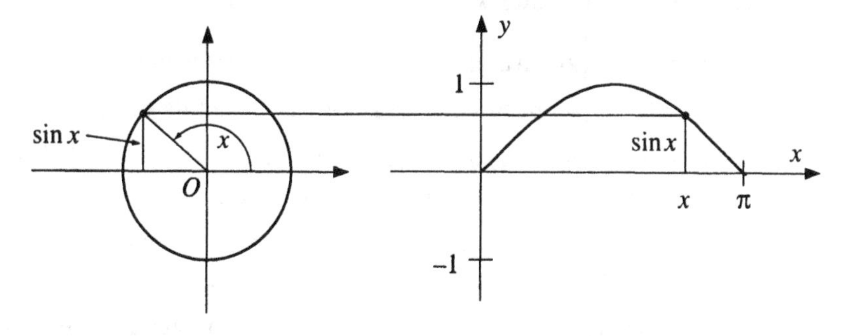

And in the third and fourth quadrants, the situation is like this:

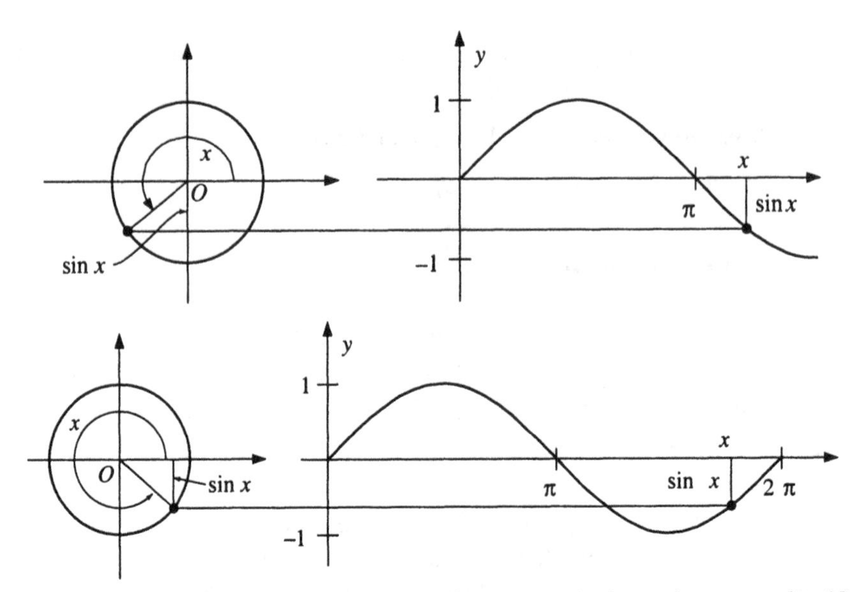

After α has rotated through 2π radians, the whole cycle repeats itself. For negative values of α, the situation is the same. Here is the complete curve:

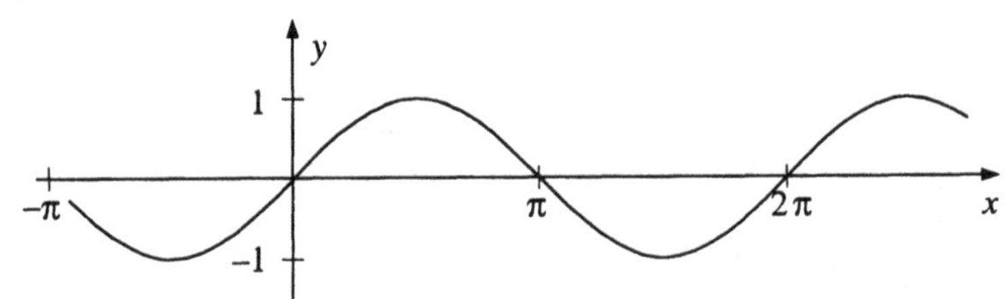

Exercise

1. Use the graph above to answer the following questions. You can check some of the answers using your calculator.

   a) Is $\sin 7\pi/5$ positive or negative? Estimate its value.

   b) Is $\sin(-3\pi/7)$ positive or negative? Estimate its value.

   c) We know that $\sin \pi/6 = 1/2$. Check this on the graph. Where else does the sine function achieve a value of $1/2$?

   d) For what values of x does $\sin x = \sin \pi/12$? Mark, on the x-axis, as many of these values as you can find.

   e) For what values of x does $\sin x = 0.8$? Estimate a value of x for which this is true. Then locate, on the x-axis, as many other values as you can find.

6 Two small miracles

We pause here to describe two remarkable relationships, so remarkable that they seem like miracles. An explanation (that is, a mathematical proof) of these miracles is postponed for later.

Miracle 1: The area under the sine curve

Look at the first arch of the curve $y = \sin x$. What can we tell about the area under this arch? The area is certainly less than π, since it fits into a rectangle whose dimensions are 1 and π:

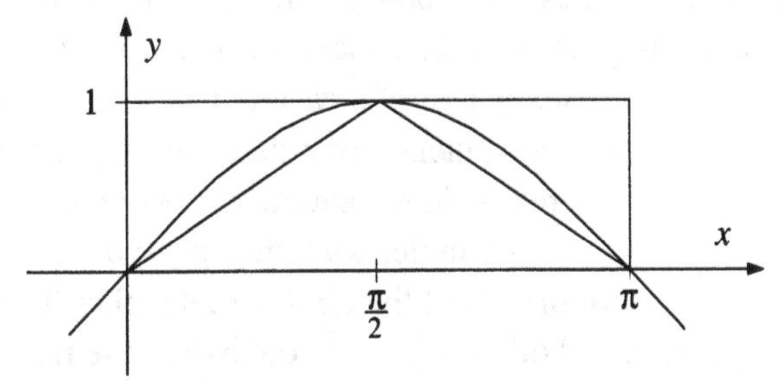

And the area is greater than that of the isosceles triangle shown in the diagram, whose area is $\pi/2$. So if we wanted to approximate the area under this arch, we would say that it is between $\pi/2$ and π. We could go further with the approximations, taking more and more triangles which would "fill" the area below the curve. Something like this is in fact done, in calculus.

The result is a small miracle: The area under one arch of the sine curve is exactly 2.

Miracle 2: The tangent to the sine curve

Let us take a point $P = (\alpha, \sin\alpha)$ on the curve $y = \sin x$. Let the perpendicular from P meet the x-axis at the point Q. Let us draw the tangent to the curve at point P, and extend it to meet the x-axis at R. It is easy to see that $PQ = \sin\alpha$.

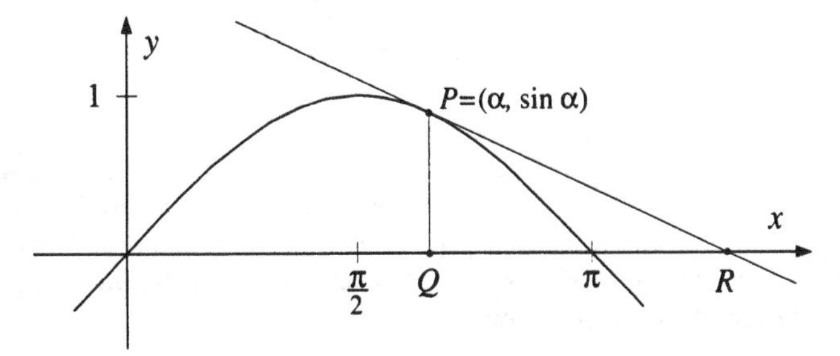

But, by a small miracle, we can also find the length of QR. It is just $|\tan\alpha|$, the absolute value of $\tan\alpha$.

Appendix – Some advantages of radian measure

Notice that the radian measure of angles, like their degree measure, is additive. That is, if two angles are placed so as to "add up" to a larger angle, the sum of the angles corresponds to the sum of the arcs.[4]

Another good thing about radian measure is that it is *dimensionless.* That is, it is independent of any unit of measurement. Length, for instance, can be measured in centimeters, inches, or miles, and we get different numbers. The same is true of area, volume, and many other quantities. But radian measure, like the sine of an angle, is a ratio, and so does not depend on the units used to measure the arc of the circle or its radius. This is another reason why physicists, and other scientists too, like to use radians.

Since the radian measure and the sine of an angle are both dimensionless, we can compare them. For an acute angle α, which is larger, $\sin\alpha$ or the radian measure of α?

Geometry can help us answer this, if the angle is small. In the diagram below, we took a circle of unit radius, and drew a tiny angle AOP. Then we made another copy of this angle (back-to-back with the first copy) and labeled it POB. Then arc AP = arc PB.

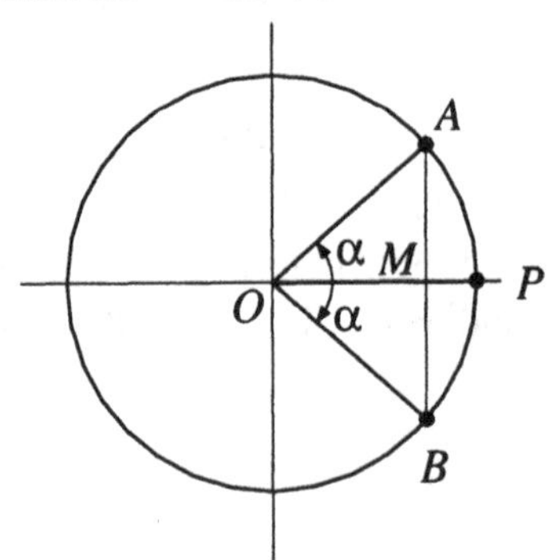

[4]Whenever we decide how to measure something, we would like the measure to be additive. Length is additive, as is area and volume. However, a trip to the grocery will quickly confirm that the price of Coca-Cola is not additive. The price of two 6-ounce bottles is likely to be more than the price of one 12-ounce bottle, because you are paying for packaging, labeling, shipping, and so on.

If the length of this arc is α, then the radian measure of $\angle AOP = \angle POB = 2\alpha/1$ (since the circle has unit radius, and $\angle AOP = \angle POB = \alpha$). From right triangle AOM, we see that $AM = \sin\alpha$, so $AB = 2\sin\alpha$. Since arc AB is longer than line segment AB, we see that $2\sin\alpha < 2\alpha$, or $\sin\alpha < \alpha$.

From this picture we also see that if the angle α is small enough, then 2α and $2\sin\alpha$ are very close to each other.[5]

Once again, we can see the advantage of radian measure. If the angle α were measured in degrees, the best statement we could make would be that $\sin\alpha < \alpha\pi/180$.

But with radian measure we can even prove a bit more. Later on we will see that for a small angle α measured in radians, the ratio $\sin\alpha/\alpha$ is very close to 1. For example, for $\alpha = 0.1$, $\sin\alpha$ is more than 99% of α itself.

Radian measure also goes well with the trigonometric ratios. We have already seen that $\sin x$ is approximately close to x for small angles. It is even closer to $x - x^3/6$, an excellent and simple approximation. We can even show that the error is less than $x^5/120$, which, for small angles, is a very tiny number.

But this is true only if we use radian measure for x. In degrees, as we have seen, this formula would be terrible.

It is true that the nicest angles have radian measures which involve the number π. And we admit that sometimes it is difficult to deal with π because it is an irrational number, and our decimal notational system for numbers doesn't provide us with a good symbol for it[6](this is why we use a Greek letter). But it's even less convenient for English-speaking people to convert miles to kilometers, or pounds to kilograms. So please don't let this slight inconvenience stop you from using radian measurement.

---

[5]What does it mean for two numbers to be "close"? For example, 1 and 0.99 are certainly close: their difference is 0.01, a tiny number. But 1000 and 998 are also close. Their difference is 2, which is a much larger number than 0.01. However, the ratio $998 : 1000 = 0.998$ is very close to 1. So sometimes we should measure "closeness" by seeing how close the *ratio* of two numbers is to 1. Thinking this way, we would not say that, 0.1 and 0.0001 are close. Although both these numbers are small, and their difference is small, their ratio is 1000, which is not small. In the diagram it is true that if α is small, not only is $\sin\alpha$ also small, but the two numbers are close, since their ratio is close to 1.

[6]The number π is one of two irrational constants that come up quite naturally. The other is e, which is approximately 2.71828, and is also irrational. The number e comes up in calculus as naturally as the number π does in geometry. About 250 years ago, it was discovered that these two numbers are related by the remarkable equation $e^{i\pi} = -1$.

Exercises

1. Use your calculator to fill in the following table (of course, the second and third columns will be numerical approximations):

α (radians)	α (degrees)	$\sin\alpha$
1	57.29578	
0.5		
0.2		
0.1		
0.01		
0.02		
0.001		
0.002		
0.005		

2. Without using your calculator, give an estimate for the value of

$$\sin 0.00123456\,.$$

   Is this estimate too large or too small? Check this using your calculator, after you've answered the question.

3. a) Use your calculator to fill in the following table:

α	$\alpha - \frac{\alpha^3}{6}$	$\sin\alpha$
1		
0.5		
0.2		
0.1		
0.05		
0.01		
0.001		

   b) The table above shows that $\sin\alpha$ is approximately equal to $\alpha - \alpha^3/6$, if α is a small angle measured in radians. Write the corresponding approximation for $\sin D$, where D is a small angle measured in degrees. Your approximation should be an expression in the variable D. Then check your expression for $D = 1°$.

4. The error in the above estimate is always less than $\alpha^5/120$. What is the largest possible error if the angle is measured in degrees, instead of radians?

5. Use your calculator to determine the radian measures of the angles x for which $x^5/120 < 0.001$.

6. We have discussed the formula

$$\sin x \approx x - \frac{x^3}{6},$$

which is proven in calculus. Can you guess the next term of this approximation?

If you can do this, you will have a formula which gives $\sin x$ for small values of x to more decimal places than most calculators can display!

7. In the year 2096, a space capsule landed on earth, with artifacts from a distant alien civilization. Here are some diagrams found in the space capsule:

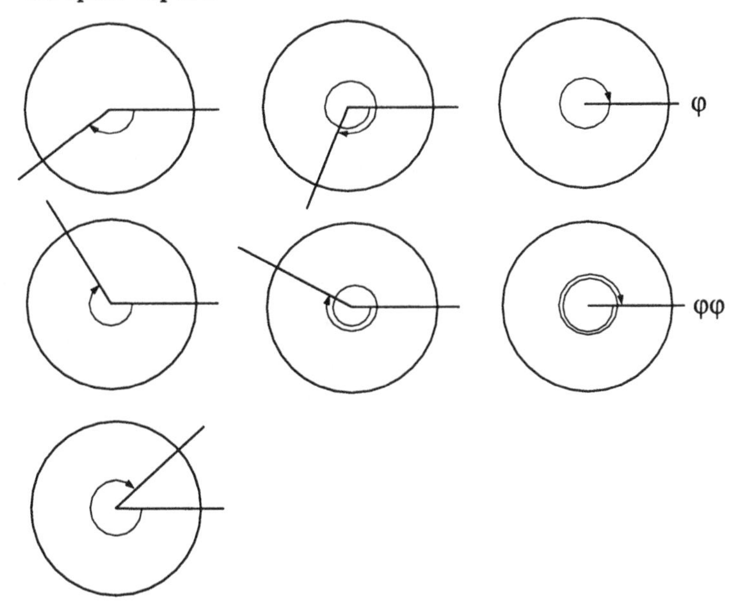

Experts believe that this chart shows how they measure angles. Tell as much as you can about the system of angle measure in this civilization. What do you think the symbol φ stands for?

8. In which quadrant do each of the following angles lie?

$$1,\ 2,\ 3,\ 4,\ 5,\ 6,\ 1000 \text{ (all these in radians)}, \quad 1000^\circ.$$

9. Suppose you answered the question above for angles of radian measure 1, 2, 3, 4, ..., 100. What fraction of these angles do you suppose would lie in quadrant 1? quadrant 2? quadrant 3? quadrant 4?

   Solution. You can guess that approximately 1/4 of the angles lie in each quadrant – there is no reason for the angles to "favor" one quadrant in particular. In fact, this guess is correct. It is a special case of the important Ergodic Theorem of higher mathematics. If you took angles of radian measure 1, 2, 3, ... up to 1000, your approximation would be even closer to 1/4 for each quadrant. □

Chapter 6

The Addition Formulas

1 More identities

We now come to an important and fundamental property of the sine and cosine functions. If we know the values of $\sin\alpha$ and $\cos\alpha$, and also the values of $\sin\beta$ and $\cos\beta$, then we can calculate the values of $\sin(\alpha+\beta)$, $\cos(\alpha+\beta)$, $\sin(\alpha-\beta)$, and $\cos(\alpha-\beta)$.

But perhaps this is easy. Perhaps $\sin(\alpha+\beta)$ is simply equal to $\sin\alpha + \sin\beta$. Let us test this guess by setting $\alpha = \beta = \pi/2$. Then $\sin(\alpha+\beta) = \sin(\pi/2+\pi/2) = \sin\pi = 0$, while $\sin\pi/2 + \sin\pi/2 = 1+1 = 2$. Since these two values are not equal, our guess is wrong.

Exercises

1. Complete the following table:

α	β	$\sin\alpha$	$\sin\beta$	$\sin\alpha+\sin\beta$	$\sin(\alpha+\beta)$
60°	30°				
$\pi/4$	$\pi/4$				
$\pi/6$	$\pi/3$				

2. Note that $\sin(\alpha+\beta)$ is not equal to $\sin\alpha+\sin\beta$ for these values of α and β. Which expression has the larger value in each case?

3. Check which of the following identities are correct, and which are not, using the angles $\alpha = 60^\circ$, $\beta = 30^\circ$:

   a) $\sin\alpha + \sin\beta = \sin(\alpha+\beta)$.

b) $\sin(\alpha - \beta) = \sin\alpha - \sin\beta$.

c) $\sin^2\alpha - \sin^2\beta = \sin(\alpha + \beta)\sin(\alpha - \beta)$.

4. a) The diagram below shows a circle with diameter $AC = 1$.

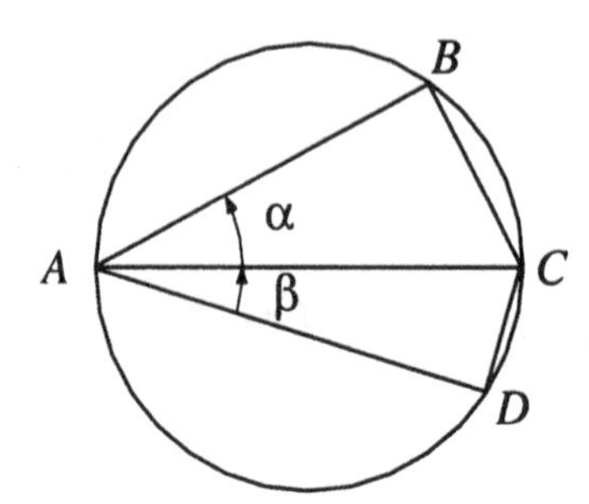

Find a line segment in the diagram equal in length to $\sin\alpha$ and one equal to $\sin\beta$.

b) The diagram below shows the same circle as above. Its diameter is still 1, but AC is not a diameter. Angles α and β are the same acute angles as before.

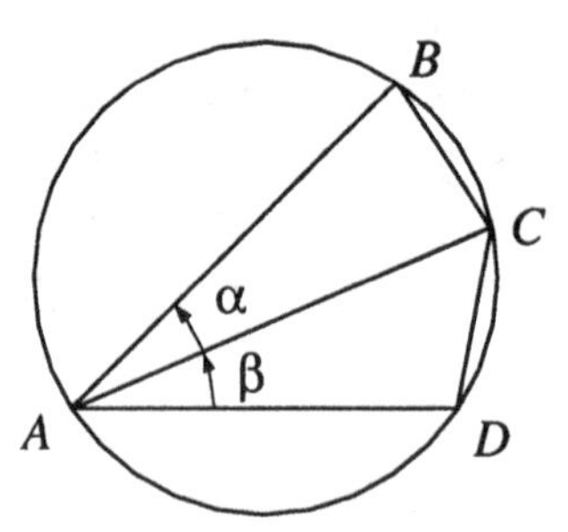

Find line segments in the diagram equal in length to $\sin\alpha$ and to $\sin\beta$.

c) In the figures for parts a) and b), draw in a line segment equal in length to $\sin(\alpha + \beta)$.

5. Recall that $\sin 45^\circ = \frac{\sqrt{2}}{2} \approx 0.707$ and $\sin 60^\circ = \frac{\sqrt{3}}{2} \approx 0.866$. Note that both these values are greater than 1/2. How can you tell immediately, without much calculation, that $\sin 45^\circ + \sin 60^\circ$ cannot equal $\sin 105^\circ$, although $45 + 60 = 105$?

2 The addition formulas

So far, most of what we have done is to give new names to familiar objects. But now we will explore the following *addition formulas* for sines and cosines:

$$\sin(\alpha+\beta) = \sin\alpha\cos\beta + \cos\alpha\sin\beta\,,$$
$$\cos(\alpha+\beta) = \cos\alpha\cos\beta - \sin\alpha\sin\beta\,.$$

In a sense, they are the key reason why the sine and cosine functions find so many uses in physics, and in mathematics as well.[1]

There are also two related formulas for differences:

$$\sin(\alpha-\beta) = \sin\alpha\cos\beta - \cos\alpha\sin\beta\,,$$
$$\cos(\alpha-\beta) = \cos\alpha\cos\beta + \sin\alpha\sin\beta\,.$$

Exercises

1. Check the formulas given above by letting $\alpha = 60°$, $\beta = 30°$.

2. Check that these formulas say something true (if not enlightening) when $\alpha = 0$ and β is any angle. What happens if $\beta = 0$?

   Note: If you ever forget which formula is which, you can quickly look at what happens if $\beta = 0$. The formula for $\sin(\alpha + 0)$, for example, should give you the value $\sin\alpha$.

3. Check the formulas for $\sin(\alpha+\beta)$ and $\cos(\alpha+\beta)$ when $\alpha+\beta = \pi/2$.

   Hint: Assume α and β are acute angles in the same triangle, and compare $\sin\alpha$ and $\cos\beta$.

4. Check that the addition formulas are true if $\alpha = \beta = \pi/4$.

5. Check that $\sin^2(\alpha+\beta) + \cos^2(\alpha+\beta) = 1$ using the formulas above. That is, show that $(\sin\alpha\cos\beta + \cos\alpha\sin\beta)^2 + (\cos\alpha\cos\beta - \sin\alpha\sin\beta)^2 = 1$.

---

[1] But these uses for sine and cosine were not the earliest. The astronomer Ptolemy, in the second century CE, used these addition formulas, although he didn't have the names sine and cosine that we use now. As an astronomer, he needed equivalent concepts to locate the stars and planets, and to describe their periodic motions.

6. Using the formulas given in the text above, prove that $\sin(\alpha+\beta)\sin(\alpha-\beta) = \sin^2\alpha - \sin^2\beta$. That is, show that

$$(\sin\alpha\cos\beta+\cos\alpha\sin\beta)(\sin\alpha\cos\beta-\cos\alpha\sin\beta) = \sin^2\alpha-\sin^2\beta\,.$$

3 Proofs of the addition formulas

The exercises above have shown that the addition and subtraction formulas we propose are reasonable, but if we are to do mathematics, we must have a proof.

We will first prove the addition formula for $\sin(\alpha+\beta)$ in the case where α, β, and $\alpha+\beta$ are all acute angles. We will need two right triangles: one containing an acute angle equal to α, and another containing an acute angle equal to β.

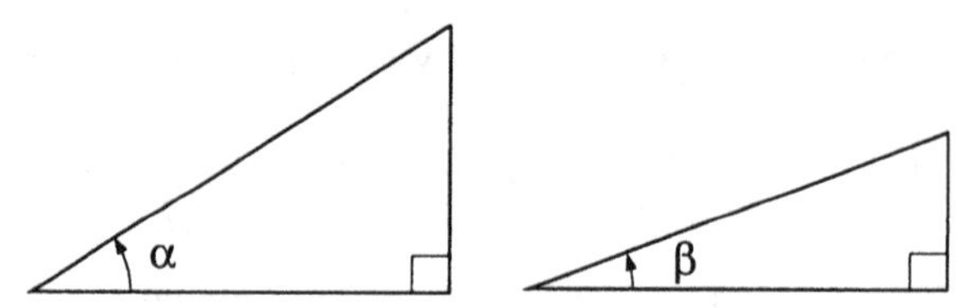

We must put these triangles together in some way, so that the resulting diagram includes an angle equal to $\alpha+\beta$ (we assumed that this angle is also acute). There are only three ways to do this, so that they have a common side:

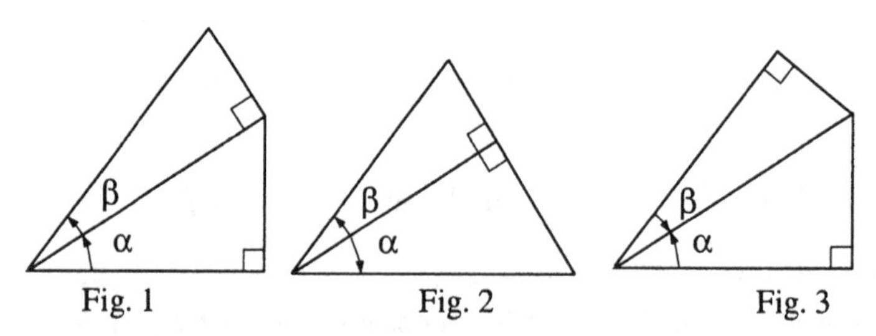

Fig. 1 Fig. 2 Fig. 3

Each of these pictures gives us a different beautiful proof of the formula for $\sin(\alpha+\beta)$. We explore here the first two. We postpone the third, which is perhaps the most interesting, for another occasion (see the appendix of this chapter).

4 A first beautiful proof

We start with Fig. 1. Let us label the sides of the triangle as shown. Then $\sin\alpha = a/c$, $\sin\beta = e/d$. We need to represent $\sin(\alpha+\beta)$ in the diagram.

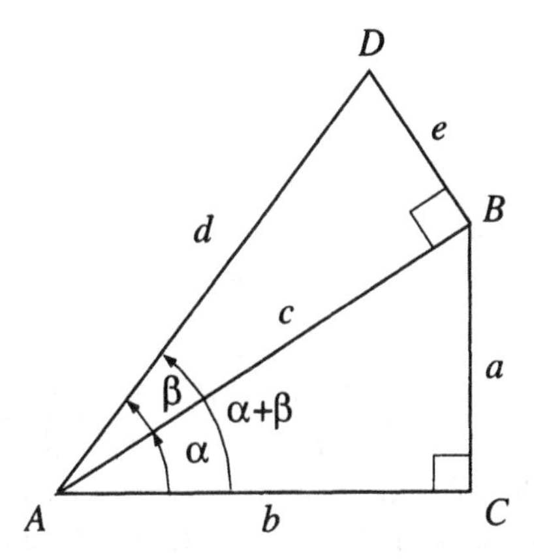

Let us draw line DQ perpendicular to the segment marked b:

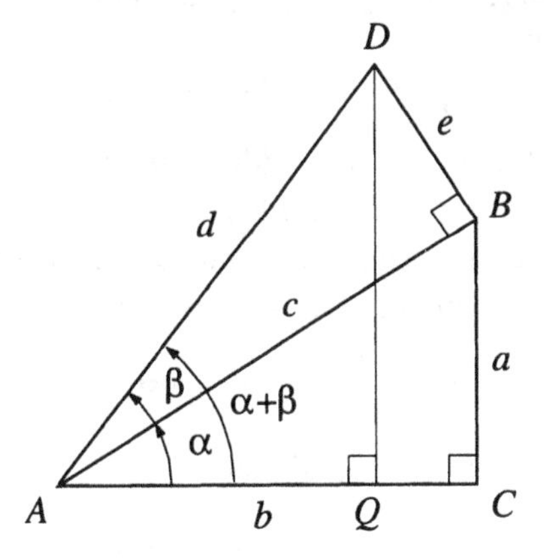

Now we can write

$$\sin(\alpha+\beta) = \frac{DQ}{d}.$$

But DQ, which is related to $\sin(\alpha+\beta)$, is not related to the ratios representing $\sin\alpha$ and $\sin\beta$. To establish this relationship, we divide DQ into two parts, with a perpendicular from point B:

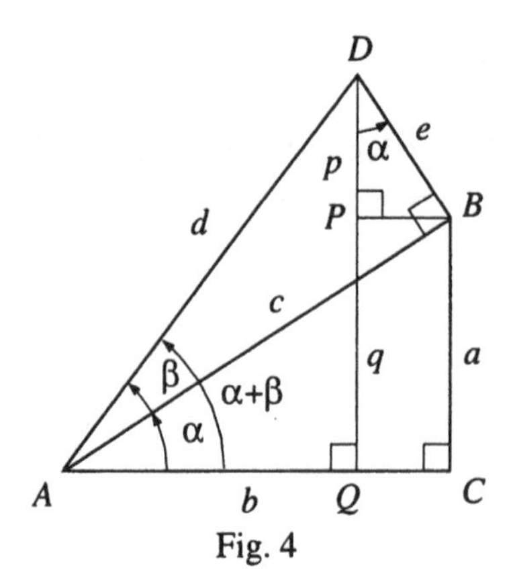

Fig. 4

Then

$$\sin(\alpha+\beta) = \frac{p+q}{d} = \frac{p}{d} + \frac{q}{d} = \frac{p}{d} + \frac{a}{d}.$$

Now we must relate p/d and a/d to $\sin\alpha$ and $\sin\beta$. We start with the second fraction. The segment a is in triangle ACB, and the segment d is in triangle ABD. We relate the fraction a/d to both triangles by introducing c as an intermediary (since c is in both triangles):

$$\frac{a}{d} = \frac{a}{c} \cdot \frac{c}{d} = \sin\alpha\cos\beta.$$

It is a bit more difficult to work with the fraction p/d. The segment d is in triangle ABD (which includes angle β), and the segment p is in triangle DPB. Happily, this last triangle contains an angle equal to α; namely[2] $\angle PDB$. Now we use segment e as an intermediary, and write

$$\frac{p}{d} = \frac{p}{e} \cdot \frac{e}{d} = \cos\alpha\sin\beta.$$

Putting this all together, we find that

$$\sin(\alpha+\beta) = \frac{a}{c}\frac{c}{d} + \frac{p}{e}\frac{e}{d} = \sin\alpha\cos\beta + \cos\alpha\sin\beta.$$

[2] If you don't see that $\angle PDB = \angle BAC = \alpha$ right away, look at the diagram below. You will see that both $\angle BAC$ and $\angle PDB$ are complementary to the angles marked γ, which are equal.

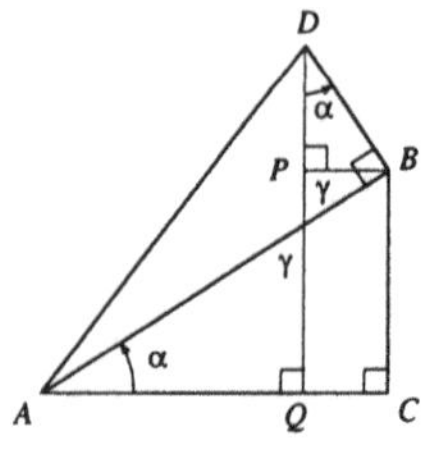

Exercises

1. We can also use the same diagram (Fig. 4) to derive a formula for $\cos(\alpha+\beta)$, where α, β, and $\alpha+\beta$ are acute angles. Let $AQ = q$, $CQ = r$. Fill in the gaps in the following proof:

$$\begin{aligned}
\cos(\alpha+\beta) &= \frac{AQ}{AD} \\
&= \frac{AC - QC}{AD} \\
&= \frac{AC}{AD} - \frac{BP}{AD} \\
&= \frac{AC}{AB}\cdot\frac{AB}{AD} - \frac{BP}{BD}\cdot\frac{BD}{AD} \\
&= \cos\alpha\cos\beta - \sin\alpha\sin\beta\,.
\end{aligned}$$

Notes:

a) Here we used two different "intermediaries": AB for the first ratio and BD for the second. Again, each intermediary plays two different roles, in two different triangles.

b) The segment QD appears with a minus sign. This is how the final formula ends up having a term subtracted rather than added.

2. Derive formulas for $\sin(\alpha-\beta)$ and $\cos(\alpha-\beta)$, using the diagram below, in terms of $\sin\alpha$, $\sin\beta$, $\cos\alpha$, and $\cos\beta$, assuming that α, β, and $\alpha-\beta$ are all positive acute angles.

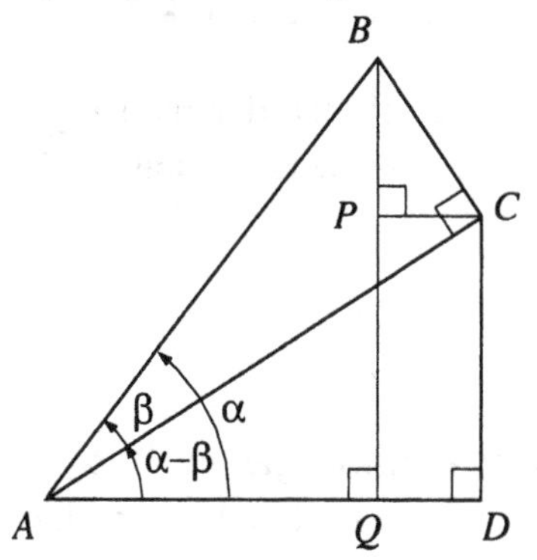

Here are the formulas to derive:

$$\sin(\alpha-\beta) = \sin\alpha\cos\beta - \cos\alpha\sin\beta\,,$$
$$\cos(\alpha-\beta) = \cos\alpha\cos\beta + \sin\alpha\sin\beta\,.$$

5 A second beautiful proof

For our second proof, we use the following theorem from Chapter 3 (see page 75).

Theorem The area of a triangle is equal to half the product of two sides and the sine of the angle between them.

In our diagram (see page 126) we have two right triangles, one including an acute angle α and the other including an acute angle β. If we place them so that they have a common side, then we get a new triangle, with one angle equal to $\alpha + \beta$:

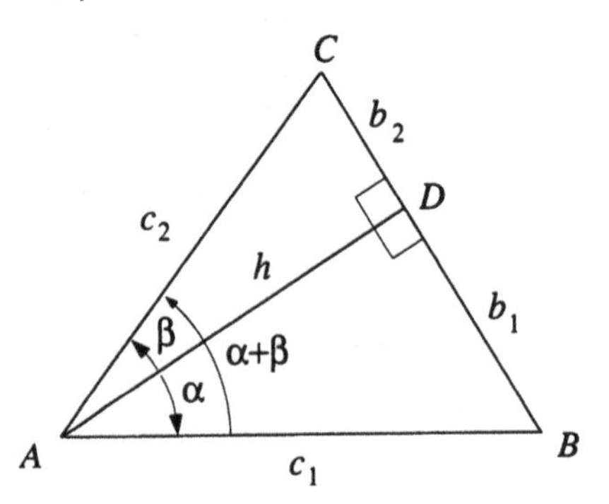

In this new triangle, the common leg of the two right triangles is an altitude, labeled h in the diagram. Each original hypotenuse is a side of the new triangle, labeled c_1 and c_2 in the diagram.

The theorem above tells us that the area of the new triangle is

$$(1/2)c_1c_2 \sin(\alpha + \beta) .$$

Let us also calculate the area of this triangle using methods of elementary geometry. The comparison of these two results will give us our formula.

Let $BD = b_1$ and $DC = b_2$. In right triangle ABD, we have

$$\frac{b_1}{c_1} = \sin \alpha ,$$

so $b_1 = c_1 \sin \alpha$. Similarly, in triangle ACD, $b_2 = c_2 \sin \beta$. Also, (from right triangle ABD), $h = c_1 \cos \alpha$, and (from right triangle ADC) $h = c_2 \cos \beta$.

Using these relationships, we can express the area of triangle ABC as:[3]

$$\begin{aligned}\tfrac{1}{2}AD\cdot BC=\tfrac{1}{2}h(b_1+b_2) &= \tfrac{1}{2}hb_1+\tfrac{1}{2}hb_2\\ &= \tfrac{1}{2}c_2\cos\beta c_1\sin\alpha+\tfrac{1}{2}c_2\cos\alpha c_1\sin\beta\,.\end{aligned}$$

Equating our two expressions for the area of triangle ABC, we have

$$\tfrac{1}{2}c_1c_2\sin(\alpha+\beta)=\tfrac{1}{2}c_1c_2\cos\beta\sin\alpha+\tfrac{1}{2}c_1c_2\cos\alpha\sin\beta\,.$$

Finally, dividing through by $(1/2)c_1c_2$ gives us the desired result. Note that $\alpha+\beta$ need not be acute for the proof to be correct (although α and β must be acute).

Exercises

1. If $\alpha = 30°$ and $\beta = 30°$, what values do our formulas give us for $\sin(\alpha+\beta)$ and $\cos(\alpha+\beta)$? Do these values agree with the values that you already know?

2. If $\sin\alpha = 3/5$ and $\sin\beta = 5/13$, what values do our formulas give us for $\sin(\alpha+\beta)$ and $\cos(\alpha+\beta)$?

3. Show that $\sin 75° = \dfrac{\sqrt{6}+\sqrt{2}}{4}$ and $\cos 75° = \dfrac{\sqrt{6}-\sqrt{2}}{4}$.

4. Find expressions in radicals (similar to those in Problem 3) for $\sin 15°$ and $\cos 15°$. Explain the coincidences.

5. a) Suppose α and β are acute angles. Can $\cos(\alpha+\beta)$ be zero?

   b) Suppose α and β are acute angles. Can $\sin(\alpha+\beta)$ be zero? Remember that neither 0 nor $\pi/2$ are considered acute.

   c) We know that if α and β are acute angles, then $\sin\alpha$, $\sin\beta$, $\cos\alpha$ and $\cos\beta$ are all positive. For acute angles α and β, must $\sin(\alpha+\beta)$ be positive? Must $\cos(\alpha+\beta)$ be positive?

6. Phoebe set out to prove the identity

$$\sin^2\alpha-\sin^2\beta=\sin(\alpha+\beta)\sin(\alpha-\beta)\,.$$

[3] Remember that the area of a triangle is half the product of any altitude and the side to which it is drawn.

She reasoned as follows:

$$\begin{aligned} \sin^2\alpha - \sin^2\beta &= (\sin\alpha + \sin\beta)(\sin\alpha - \sin\beta) \\ &= \sin(\alpha+\beta)\sin(\alpha-\beta). \end{aligned}$$

What criticism do you have of her reasoning?

7. Check the identity in Problem 6 using $\alpha = 30°$, $\beta = 60°$, on your calculator. You will find that, despite Phoebe's specious reasoning, the identity is true for these values. Is this a coincidence?

8. Prove that $\sin^2\alpha - \sin^2\beta = \sin(\alpha+\beta)\sin(\alpha-\beta)$.

9. Prove that $\cos^2\beta - \cos^2\alpha = \sin(\alpha+\beta)\sin(\alpha-\beta)$.

10. Without using your calculator, find the numerical value of $\sin 18° \cos 12° + \cos 18° \sin 12°$.

11. a) Without using your calculator, try to find the numerical value of $\sin 113° \cos 307° + \cos 113° \sin 307°$.
    b) Now use your calculator to check the result.
    c) Did you use the addition formulas in part (a)? Remember that we have proved the addition formulas only for positive acute angles. Doesn't it look like they work for larger angles as well?

12. Simplify the expression $\sin 2\alpha \cos\alpha - \cos 2\alpha \sin\alpha$.

13. Simplify the expression $\sin(\alpha+\beta)\sin\beta + \cos(\alpha+\beta)\cos\beta$.

14. Simplify the expression

$$\frac{\sin(\alpha+\beta) - \cos\alpha\sin\beta}{\cos(\alpha+\beta) + \sin\alpha\sin\beta}.$$

15. For any angle $\alpha < \pi/4$, show that

$$\sin(\alpha + \frac{\pi}{4}) = \frac{\sqrt{2}}{2}(\sin\alpha + \cos\alpha).$$

16. For any acute angles α and β for which $\cos\alpha\cos\beta \neq 0$, show that

$$\frac{\cos(\alpha+\beta)}{\cos\alpha\cos\beta} = 1 - \tan\alpha\tan\beta.$$

17. Use the law of cosines and the figure drawn for the second beautiful proof to give a direct derivation of the formula for $\cos(\alpha+\beta)$.

Appendix – Ptolemy's theorem and its connection with the addition formulas

In this appendix we explore the connection between the formula for $\sin(\alpha + \beta)$ and a remarkable geometric theorem of Ptolemy.

1. The angles of a quadrilateral inscribed in a circle

Ptolemy's theorem concerns quadrilaterals that are inscribed in circles. Suppose we have a quadrilateral $ABCD$, and we want to inscribe it in a circle. This is not always possible. In fact, if there is such a circle, then $\angle A + \angle B = \angle C + \angle D = \pi$.

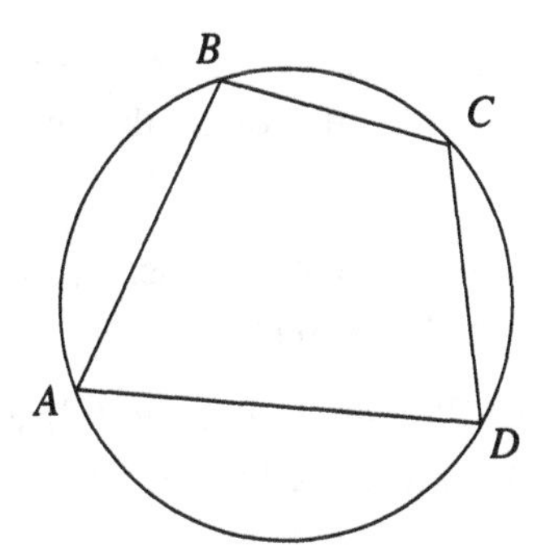

Indeed, $\angle A = \frac{1}{2}\overset{\frown}{BCD}$, and $\angle C = \frac{1}{2}\overset{\frown}{BAD}$, so $\angle A + \angle C = \frac{1}{2}(\overset{\frown}{BCD} + \overset{\frown}{BAD}) = \frac{1}{2}(2\pi) = \pi$, and similarly, $\angle B + \angle D = \pi$.

We can also show that this condition is sufficient: If the opposite angles of a quadrilateral are supplementary, then the quadrilateral can be inscribed in a circle.

To prove this, let us take some quadrilateral $ABCD$ in which $\angle B + \angle D = \pi$, and draw a circle through A, B, and C (we know that any three non-collinear points lie on a circle).

Then we can show that point D also lies on this circle. Indeed, $\angle B = \frac{1}{2}\overset{\frown}{AC}$ (the arc not containing point B), so $\overset{\frown}{ABC} = 2\pi - \overset{\frown}{AC} = 2\pi - 2\angle B$. Point D is on the circle if $\angle D = \frac{1}{2}\overset{\frown}{ABC}$ (see page 65). But in fact this is true, since $\frac{1}{2}\overset{\frown}{ABC} = \pi - \angle B = \angle D$.

So we have the following results:

Theorem A quadrilateral can be inscribed in a circle if and only if its opposite angles are supplementary.

Example 49 Suppose we want to inscribe a parallelogram in a circle. The result above tell us that its opposite angles must be supplementary, so this parallelogram must be a rectangle. Then the intersection of its diagonals will be the center of the circle, and half the diagonal will be its radius. □

2. The sides of a quadrilateral inscribed in a circle

The theorem of the last section characterizes inscribed quadrilaterals in terms of their angles. Ptolemy's theorem characterizes them in terms of the length of their sides.

A quadrilateral has four vertices, and so pairs of vertices determine six lengths. Four of these lengths are sides of the quadrilateral, and two of these lengths are diagonals. Ptolemy's theorem will use these six lengths to tell us whether or not the quadrilateral can be inscribed in a circle.

Ptolemy's Theorem A quadrilateral can be inscribed in a circle if and only if the product of its diagonals equals the sum of the products of its opposite sides.

That is, quadrilateral $ABCD$ can be inscribed in a circle if and only if $AB \times CD + AD \times BC = AC \times BD$.

Example 50 What does Ptolemy's theorem tell us for a rectangle? We know that a rectangle can be inscribed in a circle.

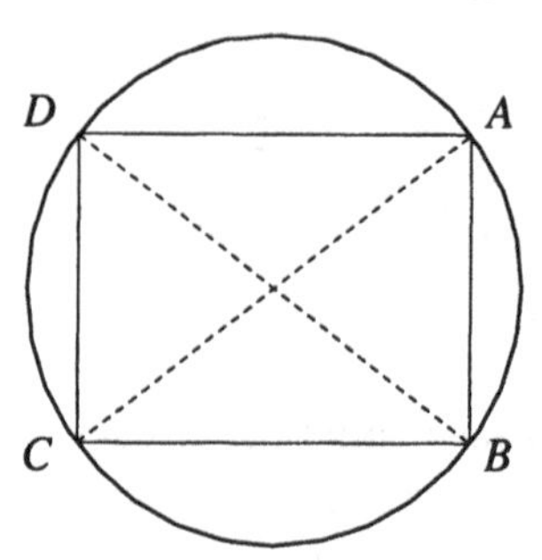

If the rectangle is $ABCD$, then we have

$$AB \times CD + AD \times BC = AC \times BD\,, \quad \text{or}$$

$$AB^2 + BC^2 = AC^2\,.$$

That is, Ptolemy's theorem here reduces to the theorem of Pythagoras.

We will not give a geometric proof of Ptolemy's theorem here. Rather, we will show that it is equivalent to the addition formula for $\sin(\alpha + \beta)$.

Ptolemy's theorem concerns the sides of a quadrilateral. Trigonometry, of course, works with angles. So our first job is to reformulate Ptolemy's theorem in terms of angles. Let us take a quadrilateral inscribed in a circle of diameter 1.

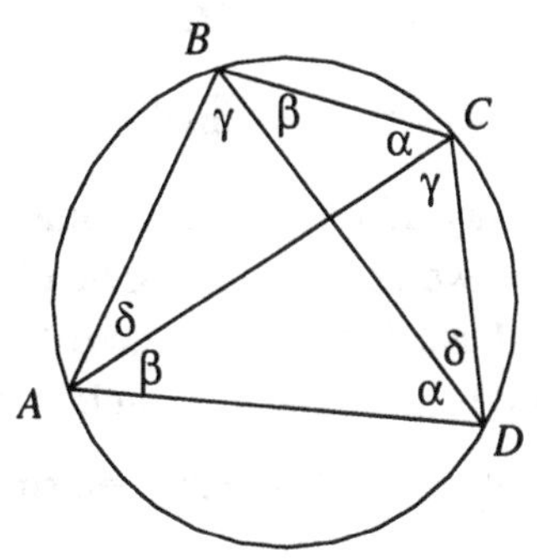

We know (Chapter 0, page 62) that in such a circle, the length of a chord is equal to the sine of its inscribed angle. If we look at the inscribed angles in the diagram, we find pairs of equal angles. These are labeled with the same Greek letter.

If we have four points A, B, C, and D, then we can divide them into pairs in three different ways:

$$\begin{array}{cc} AB & CD \\ AC & BD \\ AD & BD \end{array}$$

Each pair of points determines a length. If we take the product of these lengths, then Ptolemy's theorem says that a circle exists passing through the four points if and only if the sum of two of these products minus the third equals 0. Similarly, if we have n points, a necessary and sufficient condition that they lie on a circle is that the condition of Ptolemy's theorem is fulfilled for every choice of four of the given points.

Now we can "translate" the lengths of the quadrilateral's sides into trigonometric expressions. We have

$$AB = \sin\alpha \qquad BC = \sin\delta$$

$$CD = \sin\beta \qquad DA = \sin\gamma\ .$$

What about the diagonals? Diagonal BD is subtended by $\angle BAD$, and AC by $\angle ABC$, so we have

$$BD = \sin(\delta + \beta) = \sin(\gamma + \alpha)$$

$$AC = \sin(\alpha + \delta) = \sin(\beta + \gamma)\ .$$

Now we can write Ptolemy's theorem in trigonometric form:

$$AB \times CD + AD \times BC = AC \times BD$$

$$\sin\alpha \sin\beta + \sin\gamma \sin\delta = \sin(\beta + \gamma)\sin(\alpha + \gamma)\,.$$

Let us put this another way. If we have four angles α, β, γ, δ such that $\alpha + \beta + \gamma + \delta = \pi$, then we can divide a circle of diameter 1 into arcs of length $2\alpha, 2\beta, 2\gamma, 2\delta$ and use this circle to recreate the above figure. Since the resulting quadrilateral is inscribed in a circle, we have:

Ptolemy's Identity If $\alpha + \beta + \gamma + \delta = \pi$, then $\sin\alpha \sin\beta + \sin\gamma \sin\delta = \sin(\alpha + \gamma)\sin(\beta + \gamma)$.

This statement is equivalent to the part of Ptolemy's theorem that says that if a quadrilateral is inscribed in a circle, then the product of the diagonals equals the sum of the products of the opposite sides.

Ptolemy's theorem is a bit more general than the usual addition formula for $\sin(\alpha+\beta)$, and looks a bit nicer, since it uses only sines, and not cosines.

3. Ptolemy's identity implies the addition formula for sines

What happens to our old formula for $\sin(\alpha + \beta)$? It is a particular case of Ptolemy's identity. Indeed, suppose, in quadrilateral $ABCD$, $\alpha + \delta = \beta + \gamma = \pi/2$. Then $\sin(\beta + \gamma) = 1$, and because $\alpha + \delta = \pi/2$, we have $\sin\delta = \cos\alpha$. And since $\beta + \gamma = \pi/2$, we have $\sin\beta = \cos\gamma$. For this special case, Ptolemy's identity reduces to

$$\sin\alpha \cos\gamma + \cos\alpha \sin\gamma = 1 \cdot \sin(\alpha + \gamma)\,,$$

which is the usual addition formula. Thus Ptolemy's theorem implies Ptolemy's identity, which implies the addition formula for sines. □

4. The addition formulas imply Ptolemy's theorem

Suppose we know the addition formulas for $\sin(\alpha+\beta)$ and $\cos(\alpha+\beta)$. Let us show that we can use them to prove Ptolemy's identity.

In Ptolemy's identity, every term is the product of two sines. In order to derive this identity from the addition formulas, we need to convert these products into sums. The reader is invited to verify, using the formulas for $\cos(x \pm y)$, that

$$\sin\alpha \sin\beta = \tfrac{1}{2}\big[\cos(\alpha - \beta) - \cos(\alpha + \beta)\big]\,, \tag{1}$$

and to recall that

$$\cos(\pi + \alpha) = -\cos\alpha\,. \tag{2}$$

We use these results to prove Ptolemy's identity.

We want to show that if $\alpha + \beta + \gamma + \delta = \pi$, then

$$\sin\alpha \sin\beta + \sin\gamma \sin\delta = \sin(\beta + \gamma)\sin(\alpha + \gamma)\,.$$

The right side is a product of two sines. We use (1) to convert this to a sum of cosines:

$$\begin{aligned}\sin(\beta + \gamma)\sin(\alpha + \gamma) &= \tfrac{1}{2}\big[\cos\big((\beta + \gamma) - (\alpha + \gamma)\big) \\ &\qquad - \cos\big((\beta + \gamma) + (\alpha + \gamma)\big)\big] \\ &= \tfrac{1}{2}\big[\cos(\beta - \alpha) - \cos(\alpha + \beta + 2\gamma)\big]\,.\end{aligned}$$

We must do something about the expression $\alpha + \beta + 2\gamma$. We have $\alpha + \beta + 2\gamma = \alpha + \beta + \gamma + \delta + \gamma - \delta = \pi + \gamma - \delta$, and by (2), $\cos(\alpha + \beta + 2\gamma) = \cos(\pi + \gamma - \delta) = -\cos(\gamma - \delta)$. So if $\alpha + \beta + \gamma + \delta = \pi$, we have

$$\sin(\alpha + \beta)\sin(\alpha + \delta) = \tfrac{1}{2}\big[\cos(\beta - \alpha) + \cos(\gamma - \delta)\big]\,.$$

Since the cosine is an even function, we can write this as

$$\sin(\alpha + \beta)\sin(\alpha + \delta) = \tfrac{1}{2}\big[\cos(\alpha - \beta) + \cos(\gamma - \delta)\big]\,.$$

Now let us look at the left side of Ptolemy's identity. We have

$$\begin{aligned}\sin\alpha \sin\beta + \sin\gamma \sin\delta &= \tfrac{1}{2}\big[\cos(\alpha - \beta) - \cos(\alpha + \beta)\big] \\ &\qquad + \tfrac{1}{2}\big[\cos(\gamma - \delta) - \cos(\gamma + \delta)\big]\,.\end{aligned}$$

Now if $x + y = \pi$, we know that $\cos y = -\cos x$. Here, $(\alpha + \beta) + (\gamma + \delta) = \pi$, so $\cos(\gamma + \delta) = -\cos(\alpha + \beta)$, and we can write

$$\begin{aligned}\sin\alpha \sin\beta + \sin\gamma \sin\delta &= \tfrac{1}{2}\big[\cos(\alpha - \beta) - \cos(\alpha + \beta)\big] \\ &\qquad + \tfrac{1}{2}\big[\cos(\gamma - \delta) + \cos(\alpha + \beta)\big] \\ &= \tfrac{1}{2}\big[\cos(\alpha - \beta) + \cos(\gamma - \delta)\big]\,.\end{aligned}$$

But this is the same expression that we found equal to the left side. So Ptolemy's identity follows from the formulas for sine and cosine. □

Chapter 7

Trigonometric Identities

1 Extending the identities

Let us look back at some of our trigonometric identities. We first noted that $\sin^2\alpha + \cos^2\alpha = 1$ for any acute angle α. When we extended the definition of $\sin\alpha$ and $\cos\alpha$ to angles greater than $90°$ and less than $0°$, we noted that the identity still held true.

We have shown (in Chapter 6) that

$$\sin(\alpha + \beta) = \sin\alpha\cos\beta + \cos\alpha\sin\beta$$

for α and β positive acute angles. Is this identity still true for any angle at all?

Using the definition from Chapter 4, we can see that this formula works, for example, when $\alpha = 150°$ and $\beta = 300°$. And in fact, it will always work for angles of any size. Why is this true?

2 The Principle of Analytic Continuation: Higher mathematics to the rescue

Checking the formula for $\sin(\alpha + \beta)$ for general angles becomes very tedious. You can try it for other angles, reducing each sine or cosine to a function of a positive acute angle. But pack a lunch, because such a procedure takes a long time.

For this situation, a theorem from higher mathematics comes to our rescue. Called the Principle of Analytic Continuation, it says, roughly, that most of our identities will be preserved under the new definitions of the trigonometric functions.

More precisely, the Principle of Analytic Continuation says that *any identity involving rational trigonometric functions that is true for positive acute angles is true for any angle at all.*

Since a proof of this statement will involve results from a course in calculus and another in complex analysis, we will only state this principle here. But to understand the statement above, we must explore some terminology. A *rational trigonometric function* is a function you can get by taking the sine and cosine of various angles, together with all the constant functions, and adding, subtracting, multiplying, or dividing them.[1] Some examples of rational trigonometric functions are:

$$\frac{2\sin\alpha + 3\cos\alpha}{3\sin\alpha - 2\cos\alpha}, \qquad \sin(\alpha+\beta), \qquad \frac{2\sin\alpha + 3\cos\beta}{3\sin\alpha - 2\cos\beta},$$

$$\sin\alpha\cos\beta + \cos\alpha\sin\beta, \qquad \frac{\cos x + \sqrt{3}\sin x}{2}, \qquad \tan\alpha,$$

$$\frac{\tan\alpha + \tan\beta}{1 - \tan\alpha\tan\beta}.$$

Here are some examples which *are not* rational trigonometric functions:

$$\sqrt{\sin x}, \qquad \sqrt{1-\sin^2 x}, \qquad \sqrt{\sin^2 x - 3},$$

$$\log(\sin x), \qquad \cos(\sin x), \qquad \frac{\sqrt{\sin x}}{1-\cos x}.$$

Some of our examples should seem familiar to you. In fact, you can check that most of our identities so far have involved rational trigonometric functions.

The *Principle of Analytic Continuation* tells us that *if two such trigonometric rational functions are equal for numbers in any one interval* (*all the numbers between two real numbers*) *then they are equal for any numbers.*

For example, in our list above of rational trigonometric functions, we have the examples $\sin(\alpha+\beta)$ and $\sin\alpha\cos\beta + \cos\alpha\sin\beta$. Using geometry, we have already proved (three times!) that these two functions are equal for $0° < \alpha, \beta < 45°$ (so that α, β, and $\alpha+\beta$ are all acute angles). The Principle of Analytic Continuation says that these two functions must

[1] In the same way, if you start with integers, you can get all the *rational numbers* by adding, subtracting, multiplying, and dividing.

then be equal for any values of α and β, and not just for the ones in the interval between 0° and 45°.

Exercises

1. For each of the functions below, state whether or not it is a rational function of $\sin\alpha$:

   a) $\sqrt{2}\sin\alpha$ b) $\sqrt{2\sin\alpha}$ c) $\dfrac{1-\sin^2\alpha}{\sin\dfrac{\pi}{2}}$

   d) $\dfrac{1}{1+\dfrac{1}{\sin\alpha}}$ e) $\sqrt{1-\sin^2\alpha}$ f) $\sqrt{\dfrac{1-\cos\alpha}{2}}$.

2. Write each of the following expressions as rational functions of sines and cosines:

   a) $\tan\alpha$ b) $(1+\tan\alpha)(1-\tan\alpha)$

   c) $\dfrac{\tan\alpha+\tan\beta}{1-\tan\alpha\tan\beta}$ d) $\tan^2\alpha+\cot^2\alpha$

   e) $\tan\alpha\cot\alpha$ f) $1+\tan^2\alpha$.

3. For any angle α between 0 and $\pi/2$, we know $\cos\alpha=\sqrt{1-\sin^2\alpha}$. Does the Principle of Analytic Continuation guarantee that this statement is true for any angle? For example, is this identity correct if $\alpha=2\pi/3$?

4. For any angle α between 0 and $\pi/2$, we know $\sin^2\alpha+\cos^2\alpha=1$. Does the Principle of Analytic Continuation guarantee that this statement is true for any angle? For example, is this identity correct if $\alpha=2\pi/3$?

3 Back to our identities

You may imagine that a general statement such as the Principle of Analytic Continuation (and the full statement of this principle is even more general!) must have its roots in rather deep properties of functions. And in fact it does. This is why one needs to follow two advanced courses of mathematics before understanding it fully.

So we can continue to work with our identities, with the assurance of the mathematicians, who have proved the Principle of Analytic Continuation, that our work is valid for angles of any measure, and not just for positive acute angles.

Here, once again, are our formulas. We repeat them to emphasize their added meaning. Because of the Principle of Analytic Continuation, they are true for angles of any measure, and not just acute angles:

$$\begin{array}{|rcl|}\hline \sin(\alpha+\beta) &=& \sin\alpha\cos\beta+\cos\alpha\sin\beta \\ \sin(\alpha-\beta) &=& \sin\alpha\cos\beta-\cos\alpha\sin\beta \\ \cos(\alpha+\beta) &=& \cos\alpha\cos\beta-\sin\alpha\sin\beta \\ \cos(\alpha-\beta) &=& \cos\alpha\cos\beta+\sin\alpha\sin\beta \\ \hline \end{array}$$

Exercises

1. If α and β are acute angles such that $\sin\alpha = 3/5$ and $\sin\beta = 5/13$, find the numerical value of $\sin(\alpha+\beta)$ amd $\cos(\alpha+\beta)$. In what quadrant does the angle $\alpha+\beta$ lie?

2. If α and β are acute angles such that $\sin\alpha = 4/5$ and $\sin\beta = 12/13$, find the numerical value of $\sin(\alpha+\beta)$ and $\cos(\alpha+\beta)$. In what quadrant does the angle $\alpha+\beta$ lie?

3. If α and β are angles such that $\sin\alpha = 3/5$ and $\sin\beta = 5/13$, find $\sin(\alpha+\beta)$. (Note that we don't specify here that α and β are acute angles.) How many possible answers are there?

4. Verify that $\sin(\alpha-\beta) = \sin\alpha\cos\beta - \cos\alpha\sin\beta$ for:

   a) $\alpha = \dfrac{2\pi}{3}, \beta = \dfrac{\pi}{3}$.

   b) $\alpha = \dfrac{\pi}{4}, \beta = \dfrac{3\pi}{4}$.

   c) $\alpha = -\dfrac{\pi}{6}, \beta = \dfrac{3\pi}{2}$.

5. Show that $\cos^2\alpha + \cos^2(2\pi/3+\alpha) + \cos^2(2\pi/3-\alpha) = 3/2$.

6. Show that $\sin(x+y) + \sin(x-y) = 2\sin x\cos y$.

7. Simplify $\cos(x+y) + \cos(x-y)$.

8. Show that $\cos(x+y)\cos(x-y) = \cos^2 x \cos^2 y - \sin^2 x \sin^2 y$.

9. Show that $\sin(x+y)\sin(x-y) = \sin^2 x \cos^2 y - \cos^2 x \sin^2 y$.

10. Using the previous two exercises, show that

$$\cos(x+y)\cos(x-y) - \sin(x+y)\sin(x-y) = \cos^2 x - \sin^2 x\,.$$

Note that the left side depends on both x and y, but the right side depends only on x.

Remark We can simplify the expression

$$\cos(x+y)\cos(x-y) - \sin(x+y)\sin(x-y)$$

in another way. Let us put $A = x+y$, $B = x-y$. Then we have $\cos(x+y)\cos(x-y) + \sin(x+y)\sin(x-y) = \cos A\cos B - \sin A\sin B$. But this is just $\cos(A+B)$. However, $A+B = (x+y)+(x-y) = 2x$. Hence,

$$\cos(x+y)\cos(x-y) + \sin(x+y)\sin(x-y) = \cos 2x\,.$$

We see once more that the value of the expression

$$\cos(x+y)\cos(x-y) + \sin(x+y)\sin(x-y)$$

is independent of y.

11. We now have a slight misunderstanding. From Exercise 10 we see that the expression we are interested in equals $\cos^2 x - \sin^2 x$. And in the remark to that same exercise we see that it is equal to $\cos 2x$. Is this an error? Try to prove that it is not.

12. Show that $\cos(\alpha+\beta)\cos\beta + \sin(\alpha+\beta)\sin\beta$ does not depend on β.

4 A formula for $\tan(\alpha+\beta)$

Let us now show that $\tan(\alpha+\beta) = \dfrac{\tan\alpha + \tan\beta}{1 - \tan\alpha\tan\beta}$. We use the addition formulas to write:

$$\tan(\alpha+\beta) = \frac{\sin(\alpha+\beta)}{\cos(\alpha+\beta)} = \frac{\sin\alpha\cos\beta + \cos\alpha\sin\beta}{\cos\alpha\cos\beta - \sin\alpha\sin\beta}.$$

We now can divide the numerator and denominator by $\cos\alpha\cos\beta$:

$$\frac{\sin\alpha\cos\beta + \cos\alpha\sin\beta}{\cos\alpha\cos\beta - \sin\alpha\sin\beta} = \frac{\dfrac{\sin\alpha\cos\beta}{\cos\alpha\cos\beta} + \dfrac{\cos\alpha\sin\beta}{\cos\alpha\cos\beta}}{\dfrac{\cos\alpha\cos\beta}{\cos\alpha\cos\beta} - \dfrac{\sin\alpha\sin\beta}{\cos\alpha\cos\beta}}$$

$$= \frac{\dfrac{\sin\alpha}{\cos\alpha} + \dfrac{\sin\beta}{\cos\beta}}{1 - \dfrac{\sin\alpha\sin\beta}{\cos\alpha\cos\beta}}.$$

This leads to:

$$\boxed{\tan(\alpha+\beta) = \frac{\tan\alpha + \tan\beta}{1 - \tan\alpha\tan\beta}}$$

In a way, this is nicer than the formula for $\sin(\alpha+\beta)$ and $\cos(\alpha+\beta)$, since it uses only the tangents of α and β. The formula for $\sin(\alpha+\beta)$, on the other hand, uses $\cos\alpha$ and $\cos\beta$ as well as $\sin\alpha$ and $\sin\beta$.

Exercises

1. Check that our formula for $\tan(\alpha+\beta)$ is correct for $\alpha = 7\pi/6$, $\beta = 5\pi/3$.

2. Find a formula for $\tan(\alpha-\beta)$ in terms of $\tan\alpha$ and $\tan\beta$.

3. Show that
$$\tan\left(\frac{\pi}{4} + \alpha\right) = \frac{1+\tan\alpha}{1-\tan\alpha}.$$

4. Show that
$$\tan\left(\frac{\pi}{4} - \alpha\right) = \frac{1-\tan\alpha}{1+\tan\alpha}.$$

5. If $\alpha + \beta = \pi/4$, prove that $(1+\tan\alpha)(1+\tan\beta) = 2$.

6. Find an expression for $\tan(\alpha + \beta + \gamma)$ which involves only $\tan\alpha$, $\tan\beta$, and $\tan\gamma$.

7. Using the result of Problem 6, or otherwise, show that if $\alpha + \beta + \gamma = \pi$ (for example, if they are the three angles of a triangle), then $\tan\alpha + \tan\beta + \tan\gamma = \tan\alpha \tan\beta \tan\gamma$.

8. Show that $\tan\alpha \tan 2\alpha \tan 3\alpha = \tan 3\alpha - \tan 2\alpha - \tan\alpha$ whenever all these expressions are defined. For what values of α are some of these expressions not defined?

5 Double the angle

If we know $\sin\alpha$ and $\cos\alpha$, we can find the value of $\sin 2\alpha$ and $\cos 2\alpha$.

We know that

$$\sin(\alpha + \beta) = \sin\alpha \cos\beta + \cos\alpha \sin\beta$$

and

$$\cos(\alpha + \beta) = \cos\alpha \cos\beta - \sin\alpha \sin\beta\,.$$

Let $\alpha = \beta$. Then we have:

$$\begin{aligned} \sin 2\alpha &= \sin(\alpha + \alpha) = \sin\alpha\cos\alpha + \cos\alpha\sin\alpha \\ &= 2\sin\alpha\cos\alpha\,, \\ \cos 2\alpha &= \cos(\alpha + \alpha) = \cos\alpha\cos\alpha - \sin\alpha\sin\alpha \\ &= \cos^2\alpha - \sin^2\alpha\,. \end{aligned}$$

The formula for $\cos 2\alpha$ is particularly interesting. Since $\cos^2\alpha = 1 - \sin^2\alpha$, we can write $\cos 2\alpha = 1 - 2\sin^2\alpha$.

Similarly, since $\sin^2\alpha = 1 - \cos^2\alpha$, we can write $\cos 2\alpha = 2\cos^2\alpha - 1$. The reader is invited to check these computations.

So we have four beautiful and useful formulas:

$$\boxed{\begin{aligned} \sin 2\alpha &= 2\sin\alpha\cos\alpha \\ \cos 2\alpha &= \cos^2\alpha - \sin^2\alpha \\ \cos 2\alpha &= 2\cos^2\alpha - 1 \\ \cos 2\alpha &= 1 - 2\sin^2\alpha \end{aligned}}$$

Problem 1. If $\cos\alpha = \sqrt{3}/2$, find $\cos 2\alpha$.

Solution. We have

$$\cos 2\alpha = 2\cos^2\alpha - 1 = 2\left(\frac{\sqrt{3}}{2}\right)^2 - 1 = \frac{1}{2}.$$

(The reader should check that the other two formulas for $\cos 2\alpha$ lead to the same answer.) □

Problem 2. If $\cos\alpha = \sqrt{3}/2$, find $\sin 2\alpha$.

Solution. We have $\sin 2\alpha = 2\sin\alpha\cos\alpha$, and before we go any further we must compute $\sin\alpha$. But $\sin\alpha$ is not uniquely determined. (After all, we are not given the value of α, but only of $\cos\alpha$. More than one angle has a cosine equal to $\sqrt{3}/2$.)

To compute $\sin\alpha$, we recall that

$$\sin^2\alpha = 1 - \cos^2\alpha = 1 - \left(\frac{\sqrt{3}}{2}\right)^2 = \frac{1}{4},$$

so

$$\sin\alpha = \pm\frac{1}{2}.$$

Then

$$\sin 2\alpha = 2\sin\alpha\cos\alpha = 2\left(\pm\frac{1}{2}\right)\left(\frac{\sqrt{3}}{2}\right) = \pm\frac{\sqrt{3}}{2}.$$

The reader should check that in fact there are values of α for which each of our two answers is correct. If we are given the value of $\cos\alpha$, then the value of $\cos 2\alpha$ is determined, but the value of $\sin 2\alpha$ is not. (And certainly the value of α itself is not determined.) □

The "double angle" formulas are often used in the following form. If we write $\alpha = 2\beta$, then $\beta = \alpha/2$, and we have:

$$\sin\beta = 2\sin\frac{\beta}{2}\cos\frac{\beta}{2},$$

$$\cos\beta = \cos^2\frac{\beta}{2} - \sin^2\frac{\beta}{2} = 2\cos^2\frac{\beta}{2} - 1 = 1 - 2\sin^2\frac{\beta}{2}.$$

Exercises

1. a) If $\sin\alpha = 7/25$ and $\cos\alpha$ is positive, find $\sin 2\alpha$ and $\cos 2\alpha$.

   b) If $\sin\alpha = 7/25$ and $\cos\alpha$ is negative, find $\sin 2\alpha$ and $\cos 2\alpha$.

2. If $\sin\alpha$ and $\cos\alpha$ are both rational numbers, can $\sin 2\alpha$ be irrational? Can $\cos 2\alpha$? Check your answer with the examples given in the text, and with Exercise 1 above.

3. In doing a certain problem, a student accidentally wrote $\cos^2\alpha$ instead of $\cos 2\alpha$. But for the particular angle he was using, the answer turned out to be correct. What could these values of α have been? That is, for what values of α is $\cos^2\alpha = \cos 2\alpha$?

4. If $\sin\alpha + \cos\alpha = 0.2$, find the numerical value of $\sin 2\alpha$.

5. If $\sin\alpha - \cos\alpha = -0.3$, find the numerical value of $\sin 2\alpha$.

6. Show that $\cos 2\alpha \cos\alpha + \sin 2\alpha \sin\alpha = \cos\alpha$.

7. Show that $\sin 2\alpha \cos\alpha + \cos 2\alpha \sin\alpha = \sin 4\alpha \cos\alpha - \cos 4\alpha \sin\alpha$.

8. Prove that $\cos^2\alpha \leq \cos 2\alpha$.

9. Express $\left(\sin(\alpha/2) - \cos(\alpha/2)\right)^2$ in terms of $\sin\alpha$ only.

10. Find the numerical value of $\sin 10^\circ \sin 50^\circ \sin 70^\circ$.

    Hint: If the value of the given expression is M, find $M\cos 10^\circ$.

11. Find the numerical value of $\cos 20^\circ \cos 40^\circ \cos 80^\circ$.

12. Show that $\sin\dfrac{\pi}{10}\cos\dfrac{\pi}{5} = \dfrac{1}{4}$.

6 Triple the angle

Let us now find formulas for $\sin 3\alpha$ and $\cos 3\alpha$.

We can write

$$\begin{aligned}\sin 3\alpha &= \sin(2\alpha+\alpha)\\ &= \sin 2\alpha\cos\alpha + \cos 2\alpha\sin\alpha\\ &= 2\sin\alpha\cos^2\alpha + (1-2\sin^2\alpha)\sin\alpha\\ &= 2\sin\alpha(1-\sin^2\alpha) + (1-2\sin^2\alpha)\sin\alpha\\ &= 3\sin\alpha - 4\sin^3\alpha\,.\end{aligned}$$

(The reader should check the details.)

In the same way, we can show that

$$\cos 3\alpha = 4\cos^3\alpha - 3\cos\alpha\,.$$

Exercises

1. Complete the derivation of the formula for $\cos 3\alpha$ given above.
2. If $\sin\alpha = 3/5$, what are the possible values of $\sin 3\alpha$? Of $\cos 3\alpha$?
3. If $\cos\alpha = 4/5$, what are the possible values of $\sin 3\alpha$? Of $\cos 3\alpha$?
4. Derive formulas for $\cos 4\alpha$ in terms of (a) $\cos\alpha$ only; (b) $\sin\alpha$ only.
5. Show that $\sin 3\alpha\cos\alpha - \cos 3\alpha\sin\alpha = \sin 2\alpha$.
6. Show that
$$\frac{\sin 3\alpha}{\sin\alpha} - \frac{\cos 3\alpha}{\cos\alpha} = 2$$
for any angle α.
7. a) Show that $\sin 3\alpha = 4\sin\alpha\sin(60^\circ+\alpha)\sin(60^\circ-\alpha)$.
   b) Show that $\cos 3\alpha = 4\cos\alpha\cos(60^\circ+\alpha)\cos(60^\circ-\alpha)$.
8. Derive a formula for the ratio $\sin 4\alpha/\sin\alpha$ in terms of $\cos\alpha$.
9. Show that $\sin 3\alpha\sin^3\alpha + \cos 3\alpha\cos^3\alpha = \cos^3 2\alpha$.

7 Derivation of the formulas for $\sin\alpha/2$ and $\cos\alpha/2$

Let us now derive formulas for $\sin\alpha/2$ and $\cos\alpha/2$ in terms of trigonometric functions of α.

We being with the formula for $\cos\alpha$ in terms of $\cos\alpha/2$ (see page 146):

$$\cos\alpha = 2\cos^2\left(\frac{\alpha}{2}\right) - 1 .$$

This can be written as $2\cos^2\left(\frac{\alpha}{2}\right) = 1 + \cos\alpha$, which leads to

$$\boxed{\cos\left(\tfrac{\alpha}{2}\right) = \pm\sqrt{\frac{1+\cos\alpha}{2}}}$$

To get a formula for $\sin\alpha/2$, we proceed similarly:

$$\cos\alpha = 1 - 2\sin^2\left(\frac{\alpha}{2}\right) ,$$

or $2\sin^2\left(\frac{\alpha}{2}\right) = 1 - \cos\alpha$, which leads to

$$\boxed{\sin\left(\tfrac{\alpha}{2}\right) = \pm\sqrt{\frac{1-\cos\alpha}{2}}}$$

To show that $\tan\left(\frac{\alpha}{2}\right) = \pm\sqrt{\frac{1-\cos\alpha}{1+\cos\alpha}}$, we write

$$\tan\left(\frac{\alpha}{2}\right) = \frac{\sin\left(\frac{\alpha}{2}\right)}{\cos\left(\frac{\alpha}{2}\right)} = \frac{\pm\sqrt{\frac{1-\cos\alpha}{2}}}{\pm\sqrt{\frac{1+\cos\alpha}{2}}}$$

$$= \pm\sqrt{\frac{\frac{1-\cos\alpha}{2}}{\frac{1+\cos\alpha}{2}}}$$

$$= \pm\sqrt{\frac{1-\cos\alpha}{2}\cdot\frac{2}{1+\cos\alpha}}$$

$$= \pm\sqrt{\frac{1-\cos\alpha}{1+\cos\alpha}} .$$

In the next section, we will see two formulas for $\tan(\alpha/2)$ that are more convenient.

Exercises

1. If $\cos\alpha = 1$, find all possible values of $\cos(\alpha/2)$. You will find that there are two possible values. Give an example of a value for α which leads to each of these values.

2. Try out the formula given above for $\cos(\alpha/2)$ if

$$\text{a) } \alpha = 60^\circ, \qquad \text{b) } \alpha = 120^\circ, \qquad \text{c) } \alpha = 240^\circ.$$

   For which of these angles must we take the positive square root, and for which angles must we take the negative?

3. Fill in the following table. Note that for each of the given values of α, $\cos\alpha = \frac{1}{2}$.

α	Quadrant α?	$\alpha/2$	Quadrant $\alpha/2$?	$\cos\alpha/2$
780°				
1020°				
1140°				
1380°				
-60°				
-300°				
-420°				
-660°				
-780°				

4. Find radical expressions for $\sin 15^\circ$ and $\cos 15^\circ$.

5. Each of the "half-angle formulas" we have developed includes the square root of a trigonometric expression. Why don't we have to worry about the possibility that we are taking the square root of a negative number?

6. For positive acute angles, we can write $\cos\frac{\alpha}{2} = \sqrt{\dfrac{1+\cos\alpha}{2}}$, without the ambiguity of sign. If we could apply the Principle of Analytic Continuation to this identity, we would conclude (erroneously) that

this statement, without ambiguity of sign, was true for any angle. What is it about this identity that prevents us from applying the Principle of Analytic Continuation?

7. Suppose that the angles α, β, γ are such that $\alpha+\beta+\gamma=\pi$. (For example, α, β, γ could be the three angles of a triangle.) Show that:

   a) $\tan\frac{\alpha}{2}\tan\frac{\beta}{2}+\tan\frac{\beta}{2}\tan\frac{\gamma}{2}+\tan\frac{\gamma}{2}\tan\frac{\alpha}{2}=1.$

   Hint: Note that $\tan\frac{\alpha+\beta}{2}=\cot\frac{\gamma}{2}$, so that

   $$\tan\frac{\alpha+\beta}{2}\tan\frac{\gamma}{2}=1\,.$$

   b) $\sin\alpha+\sin\beta+\sin\gamma=4\cos\frac{\alpha}{2}\cos\frac{\beta}{2}\cos\frac{\gamma}{2}.$

8 Another formula for $\tan\alpha/2$

We showed that

$$\tan\frac{\alpha}{2}=\pm\sqrt{\frac{1-\cos\alpha}{1+\cos\alpha}}\,.$$

We can write this as

$$\begin{aligned}\tan\frac{\alpha}{2}=\pm\sqrt{\frac{1-\cos\alpha}{1+\cos\alpha}} &= \pm\sqrt{\left(\frac{1-\cos\alpha}{1+\cos\alpha}\right)\left(\frac{1+\cos\alpha}{1+\cos\alpha}\right)}\\ &= \pm\sqrt{\frac{1-\cos^2\alpha}{(1+\cos\alpha)^2}}\\ &= \pm\sqrt{\frac{\sin^2\alpha}{(1+\cos\alpha)^2}}\\ &= \pm\frac{\sin\alpha}{1+\cos\alpha}\,.\end{aligned}$$

So we have another formula for $\tan(\alpha/2)$, without radicals, but with an ambiguous sign. But in fact there is a small miracle here: we don't need the ambiguous sign! This miracle can easily be understood by looking at analytic continuation.

If the angle is positive and acute, that is, between 0° and 90°, we must select the positive sign. In other words, in this case we have

$$\tan\left(\frac{\alpha}{2}\right) = \frac{\sin\alpha}{1+\cos\alpha},$$

(without the ambiguous sign). Unlike the formula we started with, this new formula is a rational trigonometric expression, so the Principle of Analytic Continuation guarantees that in fact the equation is true for any angle.

Exercises

1. In this exercise, we check the result of the section above directly. We have shown that $\sin\alpha$ is twice the product of two particular numbers (they are $\sin\alpha/2$ and $\cos\alpha/2$), and we know that $\tan\alpha/2$ is the quotient of the same two numbers. But the product and the quotient of any two numbers always have the same sign. So $\sin\alpha$ and $\tan\alpha/2$ have the same sign. How does it now follow that

$$\tan\left(\frac{\alpha}{2}\right) = \frac{\sin\alpha}{1+\cos\alpha},$$

without ambiguity of the sign?

2. Show that

$$\tan\left(\frac{\alpha}{2}\right) = \frac{1-\cos\alpha}{\sin\alpha}.$$

9 Products to sums

We can get some further useful results by working with the formulas for $\sin(\alpha+\beta)$ and $\cos(\alpha+\beta)$. For example, we can write

$$\cos(\alpha+\beta) + \cos(\alpha-\beta) = 2\cos\alpha\cos\beta.$$

This simple yet remarkable formula says that the sum of the cosines of two angles can be written as the product of the cosines of two other angles. Perhaps this is clearer if we write it as follows:

$$\boxed{\cos\alpha\cos\beta = \tfrac{1}{2}\cos(\alpha+\beta) + \tfrac{1}{2}\cos(\alpha-\beta)}$$

So the cosine function, in a rather complicated way, "converts" products to sums. You may know that the logarithm function also "converts" products to sums, although in a much simpler fashion. In fact, people used to use cosine tables, like logarithm tables, to perform tedious multiplications by turning them into addition. If you study complex analysis you will learn of the rather deep relationship between the trigonometric functions and the exponential or logarithmic functions.

In the same way, we can write

$$\boxed{\begin{aligned} \sin\alpha \sin\beta &= \tfrac{1}{2}\cos(\alpha-\beta) - \tfrac{1}{2}\cos(\alpha+\beta) \\ \sin\alpha \cos\beta &= \tfrac{1}{2}\sin(\alpha+\beta) + \tfrac{1}{2}\sin(\alpha-\beta) \end{aligned}}$$

Exercises

1. Prove the last two identities referred to in the text.

2. Show that $\sin 75° \sin 15° = \frac{1}{4}$.

3. Show that $\sin 75° \cos 15° = \frac{2+\sqrt{3}}{4}$.

4. Find the numerical value of

$$\text{a)}\ \cos 75° \cos 15°\,, \qquad \text{b)}\ \cos 75° \sin 15°\,.$$

5. Show that

$$2\cos\left(\frac{\pi}{4}+\alpha\right)\cos\left(\frac{\pi}{4}-\alpha\right) = \cos 2\alpha\,,$$

for any angle α.

6. For any three angles α, β, γ, show that

$$\sin(\alpha+\beta)\sin(\alpha-\beta) + \sin(\beta+\gamma)\sin(\beta-\gamma) + \sin(\gamma+\alpha)\sin(\gamma-\alpha) = 0\,.$$

7. For any three angles α, β, γ, show that

$$\sin\alpha\sin(\beta-\gamma) + \sin\beta\sin(\gamma-\alpha) + \sin\gamma\sin(\alpha-\beta) = 0\,.$$

10 Sums to products

It is sometimes useful to convert sums of sines and cosines to products. The following series of examples shows how this can be done.

Example 51 Factor $\sin(\gamma + \delta) + \sin(\gamma - \delta)$.

Solution. We begin by using the addition formulas

$$\begin{aligned} \sin(\gamma + \delta) &= \sin\gamma\cos\delta + \sin\delta\cos\gamma\,, \\ \sin(\gamma - \delta) &= \sin\gamma\cos\delta - \sin\delta\cos\gamma\,. \end{aligned}$$

Adding, we find that $\sin(\gamma + \delta) + \sin(\gamma - \delta) = 2\sin\gamma\cos\delta$, which represents a factored form of the given expression. □

Example 52 A bottle and a cork together cost \$1.10. The bottle costs \$1 more than the cork. How much does the cork cost?

Solution. It is tempting to say immediately that the bottle costs \$1 and the cork costs 10 cents, but this is incorrect. With those prices, the bottle would cost only 90 cents more than the cork.

Algebra will quickly supply the correct answer. If the price of the bottle is b, and the price of the cork is c, then we have

$$\begin{aligned} b + c &= 1.1 \\ b - c &= 1\,. \end{aligned}$$

We may solve for b and c by adding these two equations. We find that $2b = 2.1$, so $b = 1.05$. Using this result, we know how to calculate the value of c from either equation. For example, using the first equation, we obtain $c = 1.1 - b = 1.1 - 1.05 = 0.05$.

Thus, the bottle costs \$1.05 and the cork costs 5 cents. □

Example 53 If $x + y = a$ and $x - y = b$, express x and y separately in terms of a and b.

Solution. Proceeding as in the problem with the bottle and the cork, we add the two equations to obtain $2x = a + b$, so $x = \frac{1}{2}(a + b)$. Then, instead of adding, we can subtract the equations, to obtain $2y = a - b$, so $y = \frac{1}{2}(a - b)$. In general, we have

If $x + y = a$ and $x - y = b$, then $x = \frac{1}{2}(a + b)$ and $y = \frac{1}{2}(a - b)$

Please remember this result. It will be useful in many applications of algebra and trigonometry, and not just in problems about bottles (of undetermined contents). □

Example 54 Write the expression $\sin\alpha + \sin\beta$ as a product of sines and cosines.

Solution. With the experience of the previous examples, this is not difficult to do. We may use Example 51 if we can find angles γ and δ such that $\gamma + \delta = \alpha$ and $\gamma - \delta = \beta$. Example 53 shows us how to do this. We just need to choose

$$\gamma = \frac{\alpha+\beta}{2}, \qquad \delta = \frac{\alpha-\beta}{2}.$$

Substituting into the result of Example 51, we obtain the useful formula

$$\boxed{\sin\alpha + \sin\beta = 2\sin\frac{\alpha+\beta}{2}\cos\frac{\alpha-\beta}{2}}$$ □

We may also express the difference $\sin\alpha - \sin\beta$ as a product of sines and cosines. We use the angles γ and δ found before, such that $\gamma + \delta = \alpha$ and $\gamma - \delta = \beta$, and write

$$\sin\alpha - \sin\beta = \sin(\gamma+\delta) - \sin(\gamma-\delta) = 2\cos\gamma\sin\delta\,.$$

We now express this result in terms of the original variables α and β, and find that

$$\boxed{\sin\alpha - \sin\beta = 2\cos\frac{\alpha+\beta}{2}\sin\frac{\alpha-\beta}{2}}$$

In the same way, we can prove the formulas

$$\boxed{\begin{aligned}\cos\alpha + \cos\beta &= 2\cos\frac{\alpha+\beta}{2}\cos\frac{\alpha-\beta}{2}\\ \cos\alpha - \cos\beta &= -2\sin\frac{\alpha+\beta}{2}\sin\frac{\alpha-\beta}{2}\end{aligned}}$$

Exercises

1. Give a detailed derivation of each of the last two formulas mentioned above.

2. Show that $\cos 70° + \sin 40° = \cos 10°$.

3. Find an acute angle α such that $\cos 55° + \cos 65° = \cos\alpha$.

4. Show that $\cos 20° + \cos 100° + \cos 140° = 0$.

5. Show that $\sin 78° + \cos 132° = \sin 18°$.

6. Show that
$$\frac{\cos 15° + \sin 15°}{\cos 15° - \sin 15°} = \sqrt{3}\,.$$

7. If $\alpha + \beta + \gamma = \pi$, show that

   a) $\sin(\alpha + \beta) = \sin\gamma$.

   b) $\cos(\alpha + \beta) = -\cos\gamma$.

   c) $\sin 2\alpha + \sin 2\beta + \sin 2\gamma = 4\sin\alpha\sin\beta\sin\gamma$.

8. For any angle α, show that
$$\sin\alpha + \sin(\alpha + 2\pi/3) + \sin(\alpha + 4\pi/3) = 0\,.$$

9. For any angle α, show that
$$\sin\alpha + 2\sin 3\alpha + \sin 5\alpha = 4\cos^2\alpha\sin 3\alpha\,.$$

10. For any three angles α, β, γ, show that
$$\frac{\sin(\beta - \gamma)}{\sin\beta\sin\gamma} + \frac{\sin(\gamma - \alpha)}{\sin\gamma\sin\alpha} + \frac{\sin(\alpha - \beta)}{\sin\alpha\sin\beta} = 0\,.$$

11. For any three angles α, β, γ, show that
$$\begin{aligned}&\sin(\alpha - \beta) + \sin(\alpha - \gamma) + \sin(\beta - \gamma)\\ &= 4\cos\frac{(\alpha - \beta)}{2}\sin\frac{(\alpha - \gamma)}{2}\cos\frac{(\beta - \gamma)}{2}\,.\end{aligned}$$

12. For any three angles α, β, γ, show that
$$\begin{aligned}&\sin(\alpha + \beta + \gamma) + \sin(\alpha - \beta - \gamma) + \sin(\alpha + \beta - \gamma)\\ &\quad + \sin(\alpha - \beta + \gamma) = 4\sin\alpha\cos\beta\cos\gamma\,.\end{aligned}$$

Appendix

I. 1. Expressions for $\sin\beta$, $\cos\beta$, and $\tan\beta$ in terms of $\tan\beta/2$.

We can use our results in trigonometry to obtain some results in number theory. Let us begin by reviewing some results obtained earlier.

Example 55 Show that $\tan^2\beta + 1 = 1/\cos^2\beta$.

Solution. $\tan^2\beta + 1 = \dfrac{\sin^2\beta}{\cos^2\beta} + 1 = \dfrac{\sin^2\beta + \cos^2\beta}{\cos^2\beta} = \dfrac{1}{\cos^2\beta}$. □

Example 56 Show that $\cos^2\beta = \dfrac{1}{1+\tan^2\beta}$.

Solution. This result follows from the previous one. □

Example 57 If $\tan\beta = a$, express in terms of a the value of $\sin 2\beta$.

Solution. We have

$$\begin{aligned}\sin 2\beta &= 2\sin\beta\cos\beta \\ &= 2\sin\beta\cos\beta\frac{\cos\beta}{\cos\beta} \\ &= \frac{2\sin\beta\cos^2\beta}{\cos\beta} \\ &= 2\tan\beta\,\cos^2\beta \\ &= \frac{2\tan\beta}{1+\tan^2\beta},\end{aligned}$$

this last because of the result of Example 56. Then, since $\tan\beta = a$, we have that

$$\sin 2\beta = \frac{2a}{1+a^2}.$$ □

In working Example 57, we have found a way to express $\sin 2\beta$ as a rational function of $\tan\beta$:

$$\sin 2\beta = \frac{2\tan\beta}{1+\tan^2\beta}.$$

Similarly, we can express $\cos 2\beta$ in terms of $\tan\beta$:

$$\cos 2\beta = \cos^2\beta - \sin^2\beta = \left(\frac{\cos^2\beta}{\cos^2\beta} - \frac{\sin^2\beta}{\cos^2\beta}\right)(\cos^2\beta)$$
$$= (1-\tan^2\beta)(\cos^2\beta)$$
$$= \frac{1-\tan^2\beta}{1+\tan^2\beta}.$$

We can also express $\tan 2\beta$ in terms of $\tan\beta$. The simplest way to do this is to use the formula we have derived for $\tan(\alpha+\beta)$:

$$\tan(\alpha+\beta) = \frac{\tan\alpha + \tan\beta}{1-\tan\alpha\tan\beta}.$$

Letting $\alpha = \beta$, we find that

$$\tan 2\beta = \frac{2\tan\beta}{1-\tan^2\beta}.$$

If we let $\tan\beta = a$, we can write

$$\boxed{\sin 2\beta = \frac{2a}{1+a^2}, \quad \cos 2\beta = \frac{1-a^2}{1+a^2}, \quad \tan 2\beta = \frac{2a}{1-a^2}}$$

which are all rational expressions in a.

Exercises

Using the above rational expressions, verify that:

1. $\sin^2\beta + \cos^2\beta = 1$
2. $\tan 2\beta = \sin 2\beta / \cos 2\beta$.

I. 2. Uniformization of $\sin\alpha$, $\cos\alpha$, and $\tan\alpha$

We can rewrite our new identities by letting $\alpha = 2\beta$:

$$\sin\alpha = \frac{2\tan\frac{\alpha}{2}}{1+\tan^2\frac{\alpha}{2}},$$

$$\cos\alpha = \frac{1-\tan^2\frac{\alpha}{2}}{1+\tan^2\frac{\alpha}{2}},$$

$$\tan\alpha = \frac{2\tan\frac{\alpha}{2}}{1-\tan^2\frac{\alpha}{2}}.$$

These formulas provide a *uniformization* of the trigonometric functions. That is, they allow us to represent all these functions using rational expressions of a single function, $\tan\alpha/2$. So, for instance, if we have a trigonometric identity, or an equation involving trigonometric functions, we can rewrite these functions as rational functions of this single variable. Then the trigonometric equation or identity becomes an algebraic equation or identity.

While this may be important theoretically, it rarely makes things easier when we have an actual problem to solve. However, this uniformization yields some very interesting results in a most unexpected area. We can use it to find *Pythagorean triples*: solutions in natural numbers to the equation $a^2 + b^2 = c^2$.

We know that if the numbers a, b, and c form a Pythagorean triple, then there is a right triangle with legs a and b and hypotenuse c. Then each acute angle of this triangle has a rational sine, cosine, and tangent. For example, we are familiar with the fact that the numbers 3, 4, and 5 satisfy the equation $a^2 + b^2 = c^2$. We can build a triangle with legs 3 and 4 and hypotenuse 5. For the smaller acute angle α of this triangle, $\sin\alpha = 3/5$, $\cos\alpha = 4/5$, and $\tan\alpha = 3/4$.

We can use our uniformization to find other triangles with angles whose sine, cosine, and tangent are rational by following this process backwards. If we let $\tan\alpha/2$ be some rational number, then our uniformization tells us that $\sin\alpha$, $\cos\alpha$, and $\tan\alpha$ will also be rational. We can then form a right triangle with rational sides and, by scaling it up, we can form a right triangle with integer sides. The sides of this triangle will be a Pythagorean triple.

For example, let

$$\tan\frac{\alpha}{2} = \frac{2}{3}.$$

Then we have

$$\sin\alpha = \frac{\frac{4}{3}}{1+\frac{4}{9}} = \frac{12}{13},$$

$$\cos\alpha = \frac{1-\frac{4}{9}}{1+\frac{4}{9}} = \frac{5}{13},$$

$$\tan\alpha = \frac{12}{5}.$$

In this case, a right triangle with an acute angle α can have sides $12/13$, $5/13$, and 1. Multiplying each side by 13, we form a similar right triangle with sides 12, 5, and 13. Because they are the sides of a right triangle, these three natural numbers satisfy the equation $a^2 + b^2 = c^2$.

Let us do this in general. Suppose

$$\tan\frac{\alpha}{2} = \frac{p}{q}.$$

Then

$$\sin\alpha = \frac{\frac{p}{q}}{1 + \frac{p^2}{q^2}} = \frac{2pq}{q^2 + p^2},$$

$$\cos\alpha = \frac{q^2 - p^2}{q^2 + p^2}.$$

Then the triangle has rational sides $2pq/(q^2+p^2)$, $(q^2-p^2)/(q^2+p^2)$, and 1, and the triangle with integer sides has sides $2pq$, $q^2 - p^2$, and $q^2 + p^2$.

Exercises

1. If $\tan(\alpha/2) = 3/2$, find the values of $\sin\alpha$, $\cos\alpha$ and $\tan\alpha$. Do these values provide us with a Pythagorean triple? with an integer right triangle?

2. What right triangle with integer sides results from letting $\tan(\alpha/2) = 5/8$ in our formulas above?

3. Verify that the numbers $2pq$, $q^2 - p^2$, and $q^2 + p^2$ satisfy the Pythagorean relationship. Which side is the hypotenuse?

II. Themes and variations

We return to a theme that we introduced in Chapter 1, and develop it more fully.

Theme: The maximum value of $\sin x \cos x$

Variation 1: Find the largest possible value of the expression $\sin x \cos x$.

Certainly $\sin x \cos x < 1$, since both $\sin x$ and $\cos x$ are at most 1 (and cannot be equal to 1 for the same angle). But is this the best estimate?

Exercises

1. With your calculator, find the value of $\sin x \cos x$ for the following values of x:

$$20^\circ, \quad 10^\circ, \quad 5^\circ \quad 1^\circ, \quad 70^\circ, \quad 80^\circ, \quad 85^\circ, \quad 89^\circ.$$

2. Without your calculator, find the value of $\sin x \cos x$ for the following values of x:

$$30^\circ, \quad 45^\circ, \quad 60^\circ.$$

Variation 2: Perhaps you have noticed some patterns in the numerical examples above. Let us see what is going on mathematically.

The product $\sin x \cos x$ reminds us of the formula $\sin 2x = 2 \sin x \cos x$. In fact, $\sin x \cos x = \sin 2x \,/\, 2$. But $\sin 2x$, like the sine of any angle, is less than 1. Hence,

$$\sin x \cos x = \frac{\sin 2x}{2} \leq \frac{1}{2}.$$

As we have seen, the value $1/2$ occurs, for example, if $x = 45^\circ$, so this is the maximum value of our expression.

Exercises

1. Find all x for which

   a) $\sin x \cos x = \frac{1}{2}$.

   b) $\sin x \cos x = \frac{\sqrt{3}}{2}$.

   c) $\sin x \cos x = \frac{\sqrt{3}}{4}$.

2. Which of the following equations has no solutions at all?

   a) $\sin x \cos x = 0.4$,
   b) $\sin x \cos x = 0.5$,
   c) $\sin x \cos x = 0.6$.

3. For what values of N does the equation

$$\sin x \cos x = N$$

   have a solution? How would you solve it?

Theme: The maximum value of $\sin x + \cos x$

Variation 1: For any x, $\sin x + \cos x < 2$, of course, since each addend on the left is at most 1 (and the addends cannot equal 1 simultaneously). Can the value be as much as $\frac{1}{2}$? Certainly: if $x = 30°$, then $\sin x = \frac{1}{2}$ and $\cos x > 0$, so $\sin x + \cos x$ is certainly greater than $\frac{1}{2}$.

Exercises

1. Check that if $x = 30°$, $\sin x + \cos x$ is greater than 1.
2. Find at least one value of x for which $\sin x + \cos x = 1$.
3. Find at least one value of x for which $\sin x + \cos x = \sqrt{2}$.

Now let us do things mathematically. Notice that $(\sin x + \cos x)^2 = \sin^2 x + \cos^2 x + 2\sin x \cos x = 1 + \sin 2x$. Since the maximum value of $\sin 2x$ is 1, the maximum value of $(\sin x + \cos x)^2 = 2$, and $\sin x + \cos x \leq \sqrt{2}$.

Exercises

1. Can $\sin x + \cos x = 1.414$?
2. Can $\sin x + \cos x = 1.415$?
3. For what values of x does $\sin x + \cos x = \sqrt{2}$?
4. What is the smallest possible value of the expression $\sin x + \cos x$? For what value of x is this minimum achieved?

Variation 2: Let us find the maximum value of $\sin x + \cos x$ in a different way, by comparing this with the formula $\sin(x + a) = \sin x \cos a + \cos x \sin a$. We can do this by using a trick. We will write

$$\sin x + \cos x = \sqrt{2}\left(\frac{1}{\sqrt{2}} \sin x + \frac{1}{\sqrt{2}} \cos x\right).$$

Why do we do this strange thing? The answer is that $\frac{1}{\sqrt{2}}$ is $\sin \frac{\pi}{4}$ and also $\cos \frac{\pi}{4}$. So we can write

$$\sin x + \cos x = \sqrt{2}\left(\sin x \cos \frac{\pi}{4} + \cos x \sin \frac{\pi}{4}\right) = \sqrt{2} \sin\left(x + \frac{\pi}{4}\right).$$

Now the largest possible value for the sine of any angle is 1, so the largest possible value for $\sin x + \cos x$ is $\sqrt{2}$.

Exercises

1. For what values of x is the maximum of $\sin x + \cos x$ achieved?

2. What is the minimum possible value of $\sin x + \cos x$? When is this minimum achieved?

Variation 3: Now let us look at the expression $3\sin x + 4\cos x$. What is its maximum value? This time, it won't help to square the quantity (try it!), so we can't use our first method.

We can compare the expression $3\sin x + 4\cos x$ to $\cos a \sin x + \sin a \cos x$. But the numbers 3 and 4 are not the sine and cosine of the same angle. However, the numbers 3 and 4 remind us of our "best friend", the 3-4-5 right triangle. In fact, the larger acute angle of this triangle has a cosine of 3/5 and a sine of 4/5. So, if we call this angle α, we can write

$$\begin{aligned} 3\sin x + 4\cos x &= 5\left(\tfrac{3}{5}\sin x + \tfrac{4}{5}\cos x\right) \\ &= 5(\cos\alpha \sin x + \sin\alpha \cos x) = 5\sin(\alpha + x). \end{aligned}$$

The maximum value of this expression is 5.

Exercises

1. In the above argument, must α be positive and acute?

2. What is the minimum value of $3\sin x + 4\cos x$? For what values of x does this occur?

3. What are the maximum and minimum values of $2\sin x + 7\cos x$?

   Hint: Take $\sqrt{53} = \sqrt{2^2 + 7^2}$, and investigate the corresponding question for $\sqrt{53}\left(\frac{2}{\sqrt{53}}\sin x + \frac{7}{\sqrt{53}}\cos x\right)$.

III. An approximation to π

We can use the half-angle formulas to find a numerical approximation to the number π.

Let us begin by checking our formulas for $\cos x/2$ and $\sin x/2$ when $x = \pi/2$. We have

$$\cos\frac{\pi/2}{2} = \cos\frac{\pi}{4} = \sqrt{\frac{1+\cos\frac{\pi}{2}}{2}} = \sqrt{\frac{1+0}{2}} = \frac{1}{\sqrt{2}} = \frac{\sqrt{2}}{2},$$

which, as we already know, is correct (note that we choose the positive sign for the radical).

Similarly, we have

$$\sin\frac{\pi/2}{2} = \sin\frac{\pi}{4} = \sqrt{\frac{1-0}{2}} = \frac{1}{\sqrt{2}} = \frac{\sqrt{2}}{2},$$

which we also expected.

Now let us get radical expressions for $\cos\pi/8$ and $\sin\pi/8$:

$$\cos\frac{\pi}{8} = \cos\frac{\pi/4}{2} = \sqrt{\frac{1+\cos\frac{\pi}{4}}{2}} = \sqrt{\frac{1+\frac{\sqrt{2}}{2}}{2}} = \sqrt{\frac{2+\sqrt{2}}{4}}$$
$$= \frac{1}{2}\sqrt{2+\sqrt{2}}.$$

$$\sin\frac{\pi}{8} = \sin\frac{\pi/4}{2} = \sqrt{\frac{1-\cos\frac{\pi}{4}}{2}} = \sqrt{\frac{1-\frac{\sqrt{2}}{2}}{2}} = \sqrt{\frac{2-\sqrt{2}}{4}}$$
$$= \frac{1}{2}\sqrt{2-\sqrt{2}}.$$

Note that the expressions we get contain "nested radicals."

Exercises

1. Finish the derivations below of radical expressions for $\cos\frac{\pi}{16}$ and $\sin\frac{\pi}{16}$:

$$\cos\frac{\pi}{16} = \cos\frac{\pi/8}{2} = \sqrt{\frac{1+\cos\frac{\pi}{8}}{2}} = \cdots = \frac{1}{2}\sqrt{2+\sqrt{2+\sqrt{2}}}.$$

$$\sin\frac{\pi}{16} = \sin\frac{\pi/8}{2} = \sqrt{\frac{1-\cos\frac{\pi}{8}}{2}} = \cdots = \frac{1}{2}\sqrt{2-\sqrt{2+\sqrt{2}}}.$$

2. Fill in the table with nested radical expressions for the values of the indicated trigonometric functions. Two of the values have been filled for you.

α	$\cos\alpha$	$\sin\alpha$
$\frac{\pi}{16}$	$\frac{1}{2}\sqrt{2+\sqrt{2+\sqrt{2}}}$	
$\frac{\pi}{32}$		$\frac{1}{2}\sqrt{2-\sqrt{2+\sqrt{2+\sqrt{2}}}}$
$\frac{\pi}{64}$		
$\frac{\pi}{128}$		

Now we know that $\cos 0 = 1$, and it is also true that the cosine of a very small angle (one whose measure is close to 0) is close to 1. The sequence of angles

$$\frac{\pi}{2},\quad \frac{\pi}{4},\quad \frac{\pi}{8},\quad \frac{\pi}{16},\quad \ldots,\quad \frac{\pi}{2^n},\quad \ldots$$

get closer and closer to 0 (approaches 0). So it is reasonable to expect that the sequence

$$\cos\frac{\pi}{2},\quad \cos\frac{\pi}{4},\quad \cos\frac{\pi}{8},\quad \cos\frac{\pi}{16},\quad \ldots,\quad \cos\frac{\pi}{2^n},\quad \ldots$$

approaches 1. In fact, this is the case. That is, the sequence:

$$0,\quad \frac{1}{2}\sqrt{2},\quad \frac{1}{2}\sqrt{2+\sqrt{2}},\quad \frac{1}{2}\sqrt{2+\sqrt{2+\sqrt{2}}},\quad \ldots$$

approaches 1. Mathematicians express this by writing

$$\lim_{n\to\infty} \frac{1}{2}\underbrace{\sqrt{2+\sqrt{2+\sqrt{2+\cdots+\sqrt{2}}}}}_{n \text{ radicals}} = 1\,.$$

Now let us look at another sequence:

$$\frac{\sin\frac{\pi}{2}}{\frac{\pi}{2}},\quad \frac{\sin\frac{\pi}{4}}{\frac{\pi}{4}},\quad \frac{\sin\frac{\pi}{8}}{\frac{\pi}{8}},\quad \ldots,\quad \frac{\sin\frac{\pi}{2^n}}{\frac{\pi}{2^n}},\quad \ldots$$

We have seen (Chapter 5) that for very small angles α, the ratio $\sin\alpha/\alpha$ is very close to 1. So the sequence above should approach 1. One way of saying this is to assert that the value $\sin\frac{\pi}{2^n}$ approaches the value $\frac{\pi}{2^n}$ (for large values of n), or that the value of

$$2^n \sin\frac{\pi}{2^n}$$

approaches π, and mathematicians have in fact proved this.

That is, they have shown (using our nested radical expressions for $\sin\frac{\pi}{2^n}$) that

$$\lim_{n\to\infty} 2^n \underbrace{\sqrt{2-\sqrt{2+\sqrt{2+\cdots+\sqrt{2}}}}}_{n\text{ radicals}} = \pi\,.$$

Exercises

1. Using your calculator or a computer, check that the expressions

$$\begin{aligned}&\tfrac{1}{2}\sqrt{2},\\ &\tfrac{1}{2}\sqrt{2+\sqrt{2}},\\ &\tfrac{1}{2}\sqrt{2+\sqrt{2+\sqrt{2}}},\\ &\tfrac{1}{2}\sqrt{2+\sqrt{2+\sqrt{2+\sqrt{2}}}},\ \ldots\end{aligned}$$

   approach 1.

2. Using your calculator or a computer, check that the expressions

$$\begin{aligned}&2^2\sqrt{2-\sqrt{2}},\\ &2^3\sqrt{2-\sqrt{2+\sqrt{2}}},\\ &2^4\sqrt{2-\sqrt{2+\sqrt{2+\sqrt{2}}}},\\ &2^5\sqrt{2-\sqrt{2+\sqrt{2+\sqrt{2+\sqrt{2}}}}},\ \ldots\end{aligned}$$

   approach π. You will have to think a bit about how to organize the computation. (The value of π is approximately $3.141592653589793\ldots$)

3. We know that

$$\cos\frac{\pi}{6} = \frac{\sqrt{3}}{2}.$$

Show that:

a) $\cos\frac{\pi}{12} = \frac{1}{2}\sqrt{2+\sqrt{3}}.$

b) $\cos\frac{\pi}{24} = \frac{1}{2}\sqrt{2+\sqrt{2+\sqrt{3}}}.$

c) $\cos\frac{\pi}{48} = \frac{1}{2}\sqrt{2+\sqrt{2+\sqrt{2+\sqrt{3}}}}.$

d) $\cos\frac{\pi}{96} = \frac{1}{2}\sqrt{2+\sqrt{2+\sqrt{2+\sqrt{2+\sqrt{3}}}}}.$

By evaluating these expressions (with a calculator or computer), observe that they are approaching 1. Can you explain why?

4. We can find another approximation to π by finding nested radical expressions for $\sin\pi/12$, $\sin\pi/24$, $\sin\pi/48$, $\sin\pi/96$, etc. Using a calculator or computer, find the values of the expressions:

a) $12\sin\frac{\pi}{12} = 6\sqrt{2-\sqrt{3}}.$

b) $24\sin\frac{\pi}{24} = 12\sqrt{2-\sqrt{2+\sqrt{3}}}.$

c) $48\sin\frac{\pi}{48} = 24\sqrt{2-\sqrt{2+\sqrt{2+\sqrt{3}}}}.$

d) $96\sin\frac{\pi}{96} = 48\sqrt{2-\sqrt{2+\sqrt{2+\sqrt{2+\sqrt{3}}}}}.$

IV. Trigonometric series

In this section we use the identities we have learned to the find the sum of series whose terms involve trigonometric expressions. This topic turns out to be of great importance in later work.

We introduce some of the techniques used by first looking at some purely algebraic problems.

Example 58 Find the sum $x + x^2 + x^3 + x^4 + \cdots + x^{100}$.

Solution. Let $S = x + x^2 + x^3 + x^4 + \cdots + x^{100}$ and multiply S by x:

$$xS = x^2 + x^3 + x^4 + \cdots + x^{101}.$$

Things get very simple if we subtract

$$\begin{aligned} S - xS = S(1-x) &= x - x^2 + x^2 - \cdots + x^{100} - x^{101} \\ &= x - x^{101}. \end{aligned}$$

Most of the terms drop out, and we find that

$$S = \frac{x - x^{101}}{1-x}. \qquad \square$$

Of course, if you already know the general formula for the sum of a geometric progression, this result is not unexpected. But if you don't already know the general formula for the sum of a geometric progression, you have essentially learned it above: the general case will work in just the same way.

The key to this trick is forming a "telescoping" sum: a sum of terms in which many pairs add up to zero.

Exercises

1. Find the sum

$$\frac{1}{\sqrt{1}+\sqrt{2}} + \frac{1}{\sqrt{2}+\sqrt{3}} + \cdots + \frac{1}{\sqrt{99}+\sqrt{100}}.$$

   Hint: Rationalize the denominators to get a telescoping sum.

2. Express in terms of n the sum $1 + 3 + 5 + \cdots + (2n+1)$.

   Hint: Write each odd integer as the difference of consecutive squares.

3. Find the product $(1+x)(1+x^2)(1+x^4)(1+x^8)(1+x^{16})$.

   Hint: Call this product P, and multiply P by $(1-x)$.

4. Without using your calculator, find the numerical value of the product $\cos 20^\circ \cos 40^\circ \cos 80^\circ$.

   Hint: Call this product P, and multiply P by the sine of a certain well-chosen angle.

V. Summing a trigonometric series

We would like to find the sum of the series

$$S = \sin x + \sin 2x + \sin 3x + \cdots + \sin nx\,.$$

We can form a telescoping sum, as in Example 58 above. The trick is to multiply by $2\sin(x/2)$:

$$2\sin\frac{x}{2}S = 2\sin\frac{x}{2}\sin x + 2\sin\frac{x}{2}\sin 2x + \cdots + 2\sin\frac{x}{2}\sin nx\,.$$

Now we turn the products into sums. The reader can recall, or check, that

$$\begin{aligned} 2\sin A\sin B &= \cos(A-B) - \cos(A+B) \\ &= \cos(B-A) - \cos(B+A)\,. \end{aligned}$$

So we can write

$$\begin{aligned} 2\sin\frac{x}{2}S &= 2\sin\frac{x}{2}\sin x + 2\sin\frac{x}{2}\sin 2x + \cdots + 2\sin\frac{x}{2}\sin nx \\ &= \left(\cos\tfrac{1}{2}x - \cos\tfrac{3}{2}x\right) + \cdots + \left(\cos(n-\tfrac{1}{2})x - \cos(n+\tfrac{1}{2})x\right) \\ &= \cos\tfrac{1}{2}x - \cos(n+\tfrac{1}{2})x\,, \end{aligned}$$

and so,

$$S = \frac{\cos\frac{1}{2}x - \cos(n+\frac{1}{2})x}{2\sin\frac{x}{2}}\,.$$

Sometimes this formula is more useful if we convert the sum in the numerator to a product. We find that

$$S = \frac{\sin\frac{n+1}{2}x\,\sin(\frac{n}{2}x)}{\sin\frac{x}{2}}\,.$$

This technique is quite general, and can be used to sum the sines or cosines of angles which are in *arithmetic progression*. We can find a general formula for the sum

$$S = \sin x + \sin(x+\alpha) + \sin(x+2\alpha) + \cdots + \sin(x+n\alpha)$$

by multiplying this sum by $2\sin\alpha/2$ and "telescoping" the result. We find that

$$S = \frac{\sin\frac{n+1}{2}\alpha\,\sin(x+\frac{n}{2}\alpha)}{\sin\frac{\alpha}{2}}\,.$$

Similarly, we can find a general formula for the sum

$$S = \cos x + \cos(x+\alpha) + \cos(x+2\alpha) + \cdots + \cos(x+n\alpha),$$

again by multiplying by $2\sin\alpha/2$, and using the identity

$$2\cos A \sin B = \sin(A+B) - \sin(A-B).$$

We find that

$$C = \frac{\sin\frac{n+1}{2}\alpha \cos(x+\frac{n}{2}\alpha)}{\sin\frac{\alpha}{2}}.$$

In the following exercises, we recommend using the hints provided, then checking the results by applying the formulas directly.

Exercises

1. Find the sum

$$\sin x + \sin 3x + \sin 5x + \cdots + \sin 99x.$$

Hint: Multiply this sum by $2\sin x$.

2. Find the sum

$$\sin x + \sin(x+\frac{\pi}{4}) + \sin(x+\frac{2\pi}{4}) + \cdots + \sin(x+\frac{99\pi}{4}).$$

Hint: Multiply this sum by $2\sin(\pi/8)$.

3. Find the sum

$$\cos 2x + \cos 4x + \cos 6x + \cdots + \cos 2nx.$$

4. Find the sum

$$\cos\frac{\pi}{k} + \cos\frac{2\pi}{k} + \cos\frac{3\pi}{k} + \cdots + \cos\frac{n\pi}{k}.$$

5. The diagram below shows a regular 24-sided polygon inscribed in a circle. A diameter of the circle is drawn, and perpendiculars are dropped from all the vertices of the polygon that lie on one side of this diameter. Find the sum of the lengths of these perpendiculars.

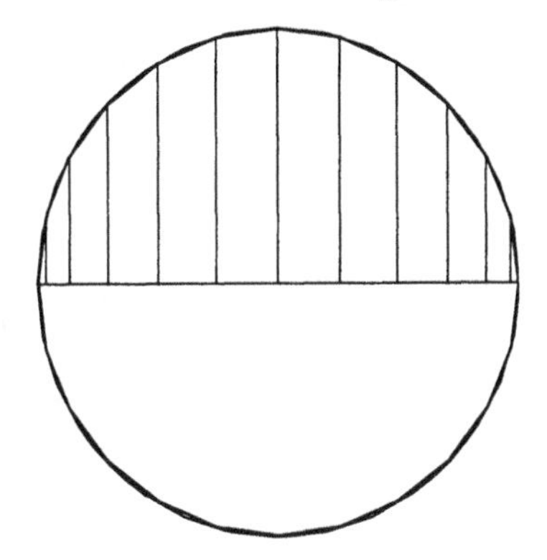

Chapter 8

Graphs of Trigonometric Functions

One of the most important uses of trigonometry is in describing periodic processes. We find many such processes in nature: the swing of a pendulum, the tidal movement of the ocean, the variation in the length of the day throughout the year, and many others.

All of these periodic motions can be described by one important family of functions, which all physicists use. These are the functions of the form

$$y = a \sin k(x - \beta)\ ,$$

where the constants a and k are positive, and β is arbitrary. In this chapter, we will describe their graphs, which we will call *sinusoidal* curves. Since they are so important, we will discuss them step-by-step, analyzing in turn each of the parameters a, k, and β.

1 Graphing the basic sine curve

$$y = a \sin k(x - \beta) \quad \text{for } a = 1, k = 1, \beta = 0$$

In Chapter 5 we drew the graph of $y = \sin x$:

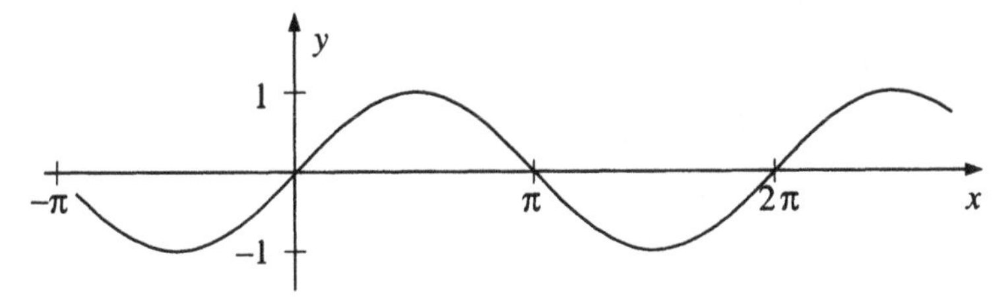

That is, we start with the case $a = 1$, $k = 1$, $\beta = 0$. Recall that we can take the sine of any real number (the *domain* of the function $y = \sin x$ is all real numbers), but that the values we get are all between -1 and 1 (the *range* of the function is the interval $-1 \leq y \leq 1$).

Let us review how we obtained this graph. On the left below is a circle with unit radius. Point P is rotating around it in a counterclockwise direction, starting at the point labeled A. If x is the length of the arc $\widehat{AP}$, then

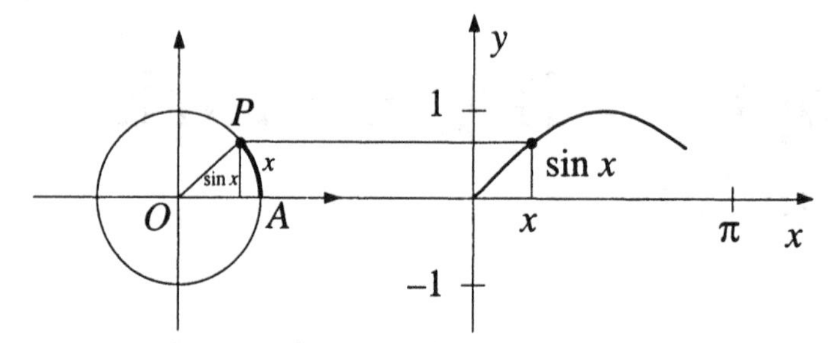

$\sin x$ is the vertical displacement of P. On the right, we have marked off the length x of arc $\widehat{AP}$. The height of the curve above the x-axis is $\sin x$.

As the angle x goes from 0 to $\pi/2$, $\sin x$ grows from 0 to 1 (the picture for $x = \pi/2$ is shown below).

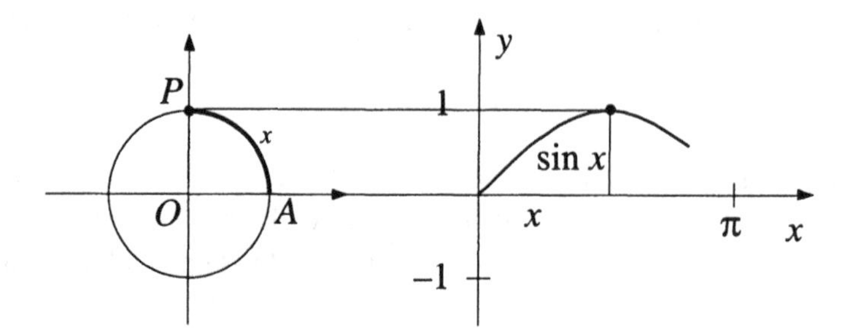

In fact, this is all we need to graph $y = \sin x$. As x goes from $\pi/2$ to π, the values of $\sin x$ repeat themselves "backwards":

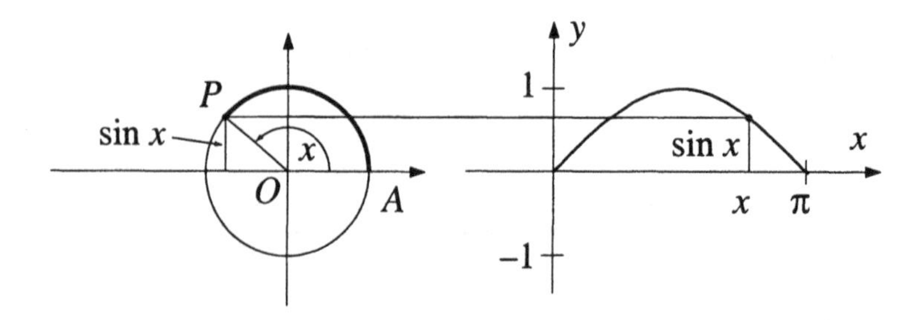

And as x goes from π to 2π, the values are the negatives of the values in the first two quadrants:

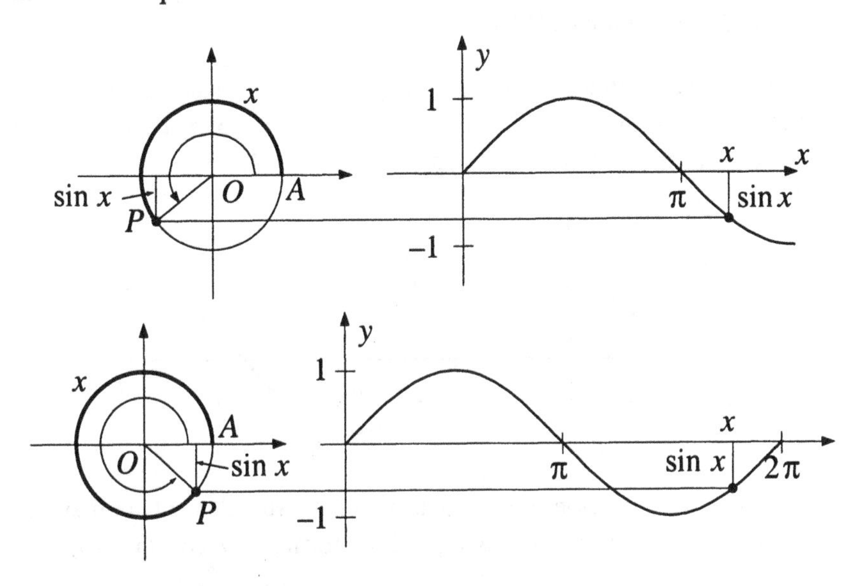

2 The period of the function $y = \sin x$

As x grows larger than 2π, the values of $\sin x$ repeat on intervals of length 2π. For this reason, we say that the function $y = \sin x$ is *periodic*, with period 2π. Geometrically, this means that if we shift the whole graph 2π units to the right or to the left, we will still have the same graph. Algebraically, this means that

$$\sin(x + 2\pi) = \sin x$$

for any number x.

Definition: A function f has a period p if $f(x) = f(x + p)$ for all values of x for which $f(x)$ and $f(x + p)$ are defined.

The function $y = \sin x$ has a period of 2π. You can check that it also has periods of $4\pi, 6\pi, -2\pi$, and in general, $2\pi n$ for any integer n. This is no accident: if $f(x)$ is a periodic function with period p, then $f(x)$ is periodic with period np for any integer n. This is why we make the following definition:

Definition: The period of a periodic function $f(x)$ is the *smallest positive* real number p such that $f(x+p) = f(x)$ for all values of x for which $f(x)$ and $f(x+p)$ are defined.

Using this definition, we say that the period of $y = \sin x$ is 2π.

Let us also draw the graph of the function $y = \cos x$. Following the same methods, we find that the graph is as shown below:

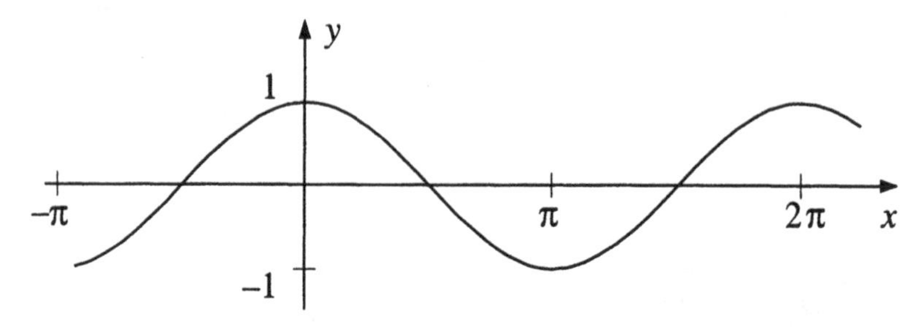

The period of the function $y = \cos x$ is also 2π. We will see later that this curve can be described by an equation of the form $y = a \sin k(x - \beta)$.

3 Periods of other sinusoidal curves

$$y = a \sin k(x - \beta) \quad \text{for } a = 1,\ \beta = 0,\ k > 0$$

Example 59 Find the period of the function $y = \sin 3x$.

Solution: One period of this function is $2\pi/3$, since $\sin 3(x + 2\pi/3) = \sin(3x + 2\pi) = \sin 3x$. It is not difficult to see that this is the smallest positive period (for example, by looking at the values of x for which $\sin 3x = 0$).

Example 60 Draw the graph of the function $y = \sin 3x$.

Solution: The function $y = \sin x$ takes on certain values as x goes from 0 to 2π. The function $y = \sin 3x$ takes on these same values, but as x goes from 0 to $2\pi/3$. Hence one period of the graph looks like this:

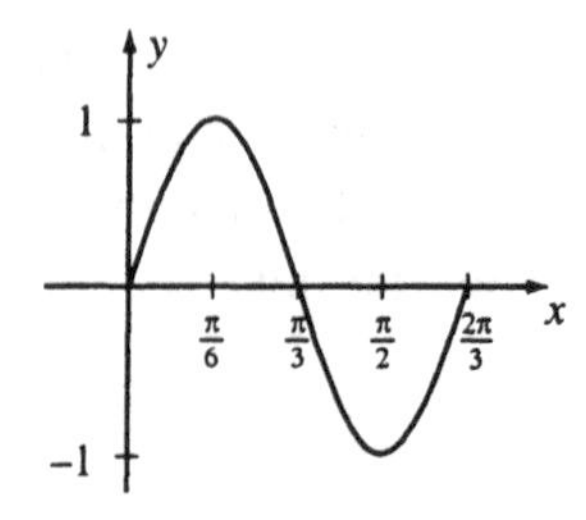

Having drawn one period, of course, it is easy to draw as much of the whole graph as we like (or have room for):

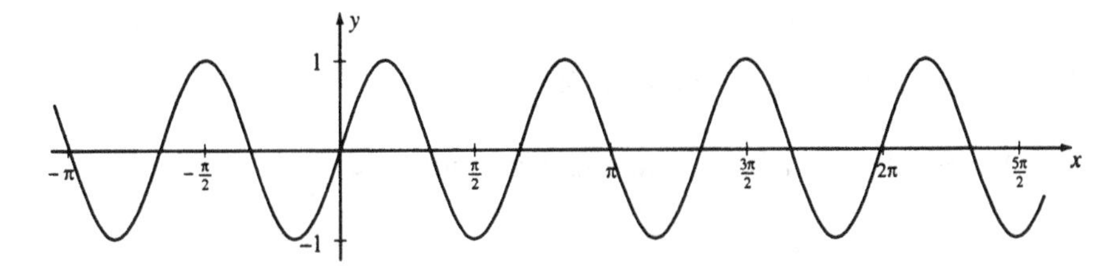

The graph is the same as that of y = sin x, but compressed by a factor of 3 in the x- direction. In general, we have the following result:

> For $k > 1$, the graph of $y = \sin kx$ is obtained from the graph $y = \sin x$ by compressing it in the x-direction by a factor of k.

What if $0 < k < 1$? Let us draw the graph of $y = \sin x/5$. Since the period of $y = \sin x/5$ is 10π, our function takes on the same values as the function $y = \sin x$, but stretched out over a longer period.

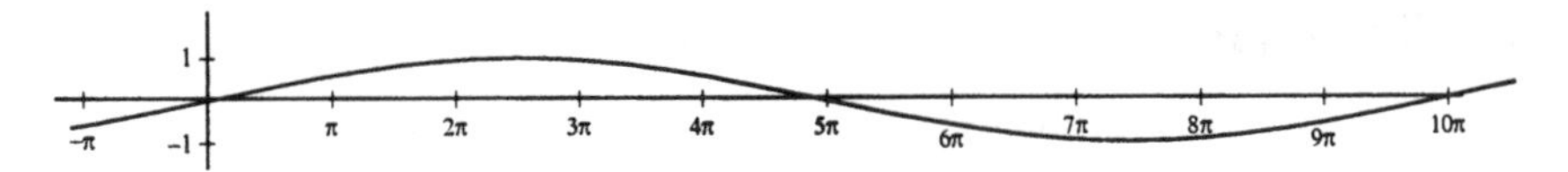

Again, we have a general result:

> For $0 < k < 1$, the graph of $y = \sin kx$ is obtained from the graph $y = \sin x$ by stretching it in the x-direction by a factor of k.

Analogous results hold for graphs of the functions $y = \cos kx$, $k > 0$.

Our basic family of functions is $y = a \sin k(x - \pi)$. What is the significance of the constant k here? We have seen that $2\pi/k$ is the period of the function. So in an interval of 2π, the function repeats its period k times. For this reason, the constant k is called the *frequency* of the function.

Exercises

Find the period and frequency of the following functions:

1. $y = \sin 5x$ 2. $y = \sin x/4$ 3. $y = \cos 4x/5$ 4. $y = \cos 5x/4$.

Graph each of the following curves. Indicate the period of each. Check your work with a graphing calculator, if you wish.

5. $y = \sin 3x$ 6. $y = \sin x/3$ 7. $y = \sin 3x/2$ 8. $y = \sin 2x/3$
9. $y = \cos 2x/3$ 10. $y = \cos 3x/2$

11. The graph shown below has some equation $y = f(x)$.

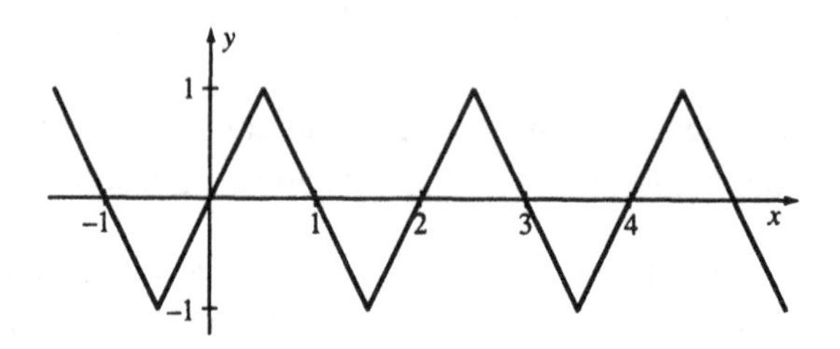

(a) Draw the graph of the function $y = f(3x)$.

(b) Draw the graph of the function $y = f(x/3)$.

4 The amplitude of a sinusoidal curve

$$y = a \sin k(x - \beta); \quad a > 0, \beta = 0, k > 0$$

Example 61 Draw the graph of the function $y = 3 \sin x$.

Solution: The values of this function are three times the corresponding values of the function $y = \sin x$. Hence the graph will have the same period, but each y-value will be multiplied by 3:

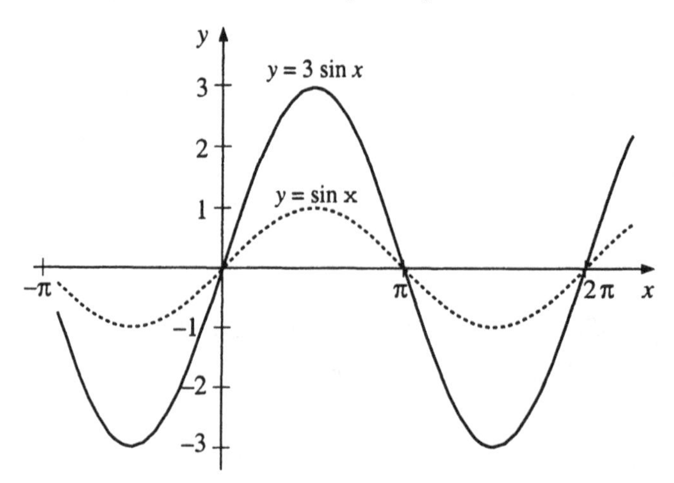

We see that the graph of $y = 3\sin x$ is obtained from the graph of $y = \sin x$ by stretching in the y-direction. Similarly, it is not hard to see that the graph of $y = (1/2)\sin x$ is obtained from the graph of $y = \sin x$ by a compression in the y-direction.

We have the following general result:

> *For $a > 1$, the graph of $y = a\sin x$ is obtained from the graph $y = \sin x$ by stretching in the y-direction. For $0 < a < 1$, the graph of $y = a\sin x$ is obtained from the graph $y = \sin x$ by compressing in the y-direction.*

Analogous results hold for graphs of functions in which the period is not 1, and for equations of the form $y = a\cos x$. The constant a is called the *amplitude* of the function $y = a\sin k(x - \beta)$.

Exercises

Graph the following functions. Give the period and amplitude of each. As usual, you are invited to check your work, after doing it manually, with a graphing calculator.

1. $y = 2\sin x$
2. $y = (1/2)\sin x$
3. $y = 3\sin 2x$
4. $y = (1/2)\sin 3x$
5. $y = 4\cos x$
6. $y = (1/3)\cos 2x$

7. Suppose $y = f(x)$ is the function whose graph is given in Exercise 11 on page 178.

   (a) Draw the graph of the function $y = 3f(x)$.

   (b) Draw the graph of the function $y = (1/3)f(x)$.

5 Shifting the sine

$$y = a\sin k(x - \beta); \quad a = 1, k = 1, \beta \text{ arbitrary}$$

We start with two examples, one in which β is positive and another in which β is negative.

Example 62 Draw the graph of the function $y = \sin(x - \pi/5)$.

Solution: We will graph this function by relating the new graph to the graph of $y = \sin x$. The positions of three particular points[1] on the original graph will help us understand how to do this:

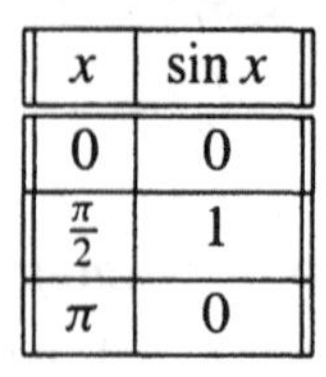

x	$\sin x$
0	0
$\frac{\pi}{2}$	1
π	0

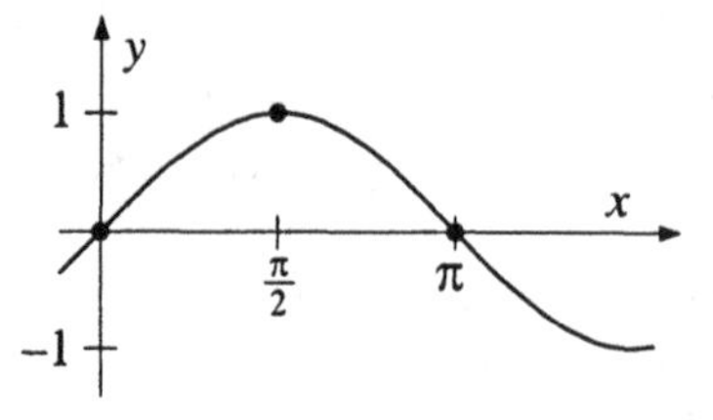

What are the analogous points on the graph of $y = \sin\left(x - \frac{\pi}{5}\right)$? It is not convenient to use $x = 0$, because then $y = \sin\left(-\frac{\pi}{5}\right)$, whose value is difficult to work with. Similarly, if we use $x = \frac{\pi}{2}$, we will need the value $y = \sin\left(\frac{\pi}{2} - \frac{\pi}{5}\right) = \sin\frac{3\pi}{10}$, which is still less convenient.

But if we let $x = \frac{\pi}{5}, \frac{\pi}{2} + \frac{\pi}{5}, \pi + \frac{\pi}{5}$, things will work out better:

x	$x - \frac{\pi}{5}$	$\sin\left(x - \frac{\pi}{5}\right)$
$\frac{\pi}{5}$	0	0
$\frac{\pi}{2} + \frac{\pi}{5}$	$\frac{\pi}{2}$	1
$\pi + \frac{\pi}{5}$	π	0

That is, our choice of "analogous" points in our new function are those where the y-values are the same as those of the original function, not where the x-values are the same. The graph of $y = \sin\left(x - \frac{\pi}{5}\right)$ looks just like the graph of $y = \sin x$, but shifted to the right by $\frac{\pi}{5}$ units:

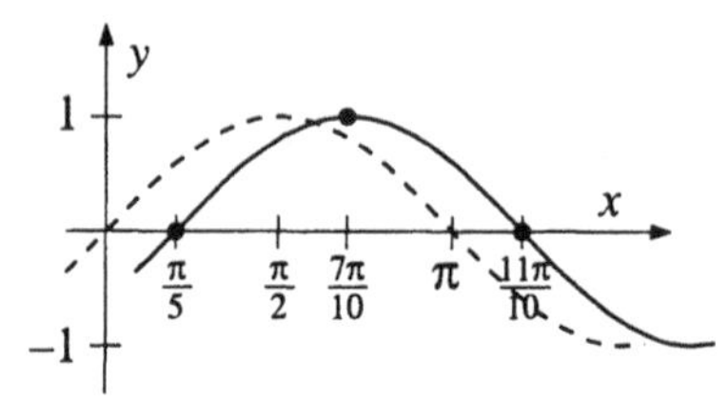

But we must check this graph for more than three points. Are the other points on the graph shifted the same way? Let us take any point $(x_0, \sin x_0)$ on the graph $y = \sin x$. If we shift it to the right by $\frac{\pi}{5}$, we are merely adding this number to the point's x-coordinate, while leaving its y-coordinate the same. The new point we obtain is $\left(x_0 + \frac{\pi}{5}, \sin x_0\right)$, and this is in fact on the graph of the function $y = \sin\left(x - \frac{\pi}{5}\right)$.

[1] Of course, with a calculator or a table of sines, you can get many more values. Or, if you have a good memory, you can remember the values of the sines of other particular angles. But these three points will serve us well for quite a while.

There is nothing special about the number $\frac{\pi}{5}$, except that it is positive. In general, the following statement is useful:

> *If $\beta > 0$, the graph of $y = \sin(x - \beta)$ is obtained from the graph of $y = \sin x$ by a shift of β units to the right.*

What if β is negative?

Example 63 Draw the graph of the function $y = \sin(x + \frac{\pi}{5})$.

Solution: In this example, $\beta = -\frac{\pi}{5}$. Again, we will relate this graph to the graph of $y = \sin x$. Using the method of the previous example, we seek values of x such that

$$\sin(x + \frac{\pi}{5}) = 0, \quad \sin(x + \frac{\pi}{5}) = 1, \quad \sin(x + \frac{\pi}{5}) = 0 \text{ (for a second time)}.$$

It is not difficult to see that these values are $x = -\frac{\pi}{5}, \frac{\pi}{2} - \frac{\pi}{5}, \pi - \frac{\pi}{5}$, respectively. Using these values, we find that the graph of $y = \sin(x + \frac{\pi}{5})$ is obtained by shifting the graph of $y = \sin x$ by $\frac{\pi}{5}$ units to the *left*:

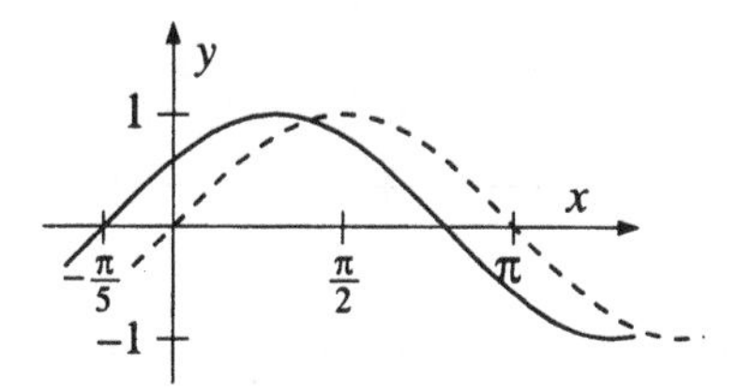

In general:

> *The graph of the function $y = \sin(x - \beta)$ is obtained from the graph of $y = \sin x$ by a shift of β units. The shift is towards the left if β is negative, and towards the right if β is positive.*

The number β is called the *phase angle* or *phase shift* of the curve. Analogous results hold for the graph of $y = \cos(x - \beta)$.

Exercises

Sketch the graphs of the following functions:

1. $y = \sin(x - \frac{\pi}{6})$
2. $y = \sin(x + \frac{\pi}{6})$
3. $y = 2\sin(x - \frac{\pi}{2})$
4. $y = \frac{1}{2}\sin(x + \frac{\pi}{2})$
5. $y = \cos(x - \frac{\pi}{4})$
6. $y = 3\cos(x + \frac{\pi}{3})$
7. $y = \sin(x - 2\pi)$

8–11: Write equations of the form $y = \sin(x - \alpha)$ for each of the curves shown below:

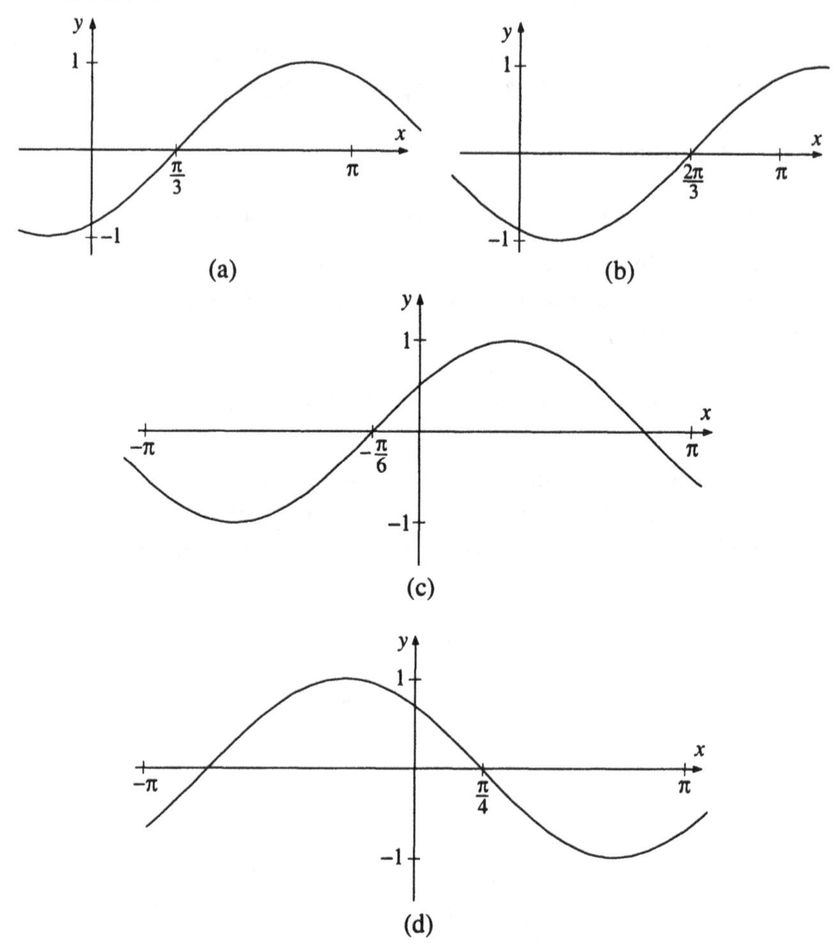

6 Shifting and stretching

Graphing $y = a \sin k(x - \beta)$

We run into a small difficulty if we combine a shift of the curve with a change in period.

Example 64 Graph the function $y = \sin(2x + \pi/3)$.

Solution: Let us write this equation in our standard form:

$$\sin(2x + \pi/3) = \sin 2(x + \pi/6)$$

We see that the graph is that of $y = \sin 2x$, shifted $\pi/6$ units to the left.

At first glance, one might have thought that the shift is $\pi/3$ units to the left. But this is incorrect. In the original equation, $\pi/3$ is added to $2x$, not to x. The error is avoided if we rewrite the equation in standard form.

Exercises

Graph the following functions:

1. $y = \sin \frac{1}{2}(x - \frac{\pi}{6})$ 2. $y = \sin(\frac{1}{2}x - \frac{\pi}{6})$ 3. $y = \cos 2(x + \frac{\pi}{3})$

4–5: Write equations of the form $y = \sin k(x - \beta)$ for the following graphs:

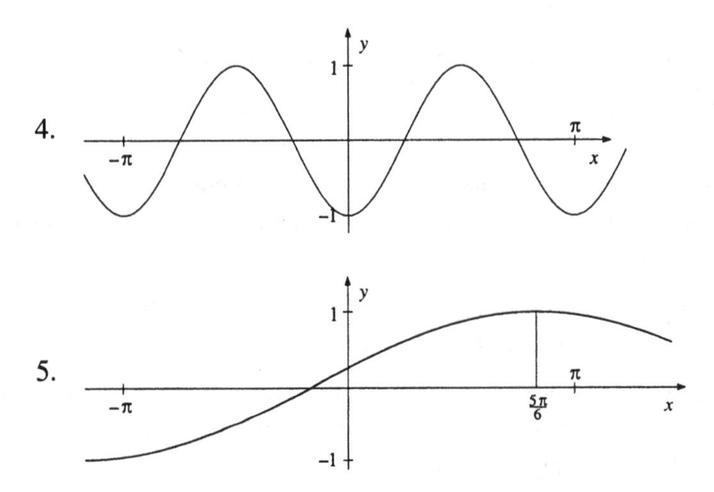

7 Some special shifts: Half-periods

We will see, in this section, that we have not lost generality by restricting a and k to be positive, or by neglecting the cosine function.

It is useful to write our general equation as $y = a \sin k(x + \gamma)$, where $\gamma = -\beta$. Then, for positive values of γ, we are shifting to the left. For the special value $c = 2\pi$, we already know what happens to the graph $y = \sin x$. Since 2π is a period of the function, the graph will coincide with itself after such a shift.

In fact, we can state the following alternative definition of a period of a function:

> *A function $y = f(x)$ has period p if the graph of the function coincides with itself after a shift to the left of p units.*

Our original definition said that a function $f(x)$ is periodic with period p if $f(x) = f(x+p)$ for all values of x for which these expressions are defined. Our new definition is equivalent to the earlier one, since the graph of $y = f(x)$, when shifted to the left by p units, is just the graph $y = f(x+p)$. These graphs are the the same if and only if $f(x) = f(x+p)$.

Let us see what happens when we shift the graph $y = \sin x$ to the left by $n\pi/2$, where n is an integer.

For $n = 1$, we have the graph $y = \sin(x + \pi/2)$, a shift to the left of the graph $y = \sin x$:

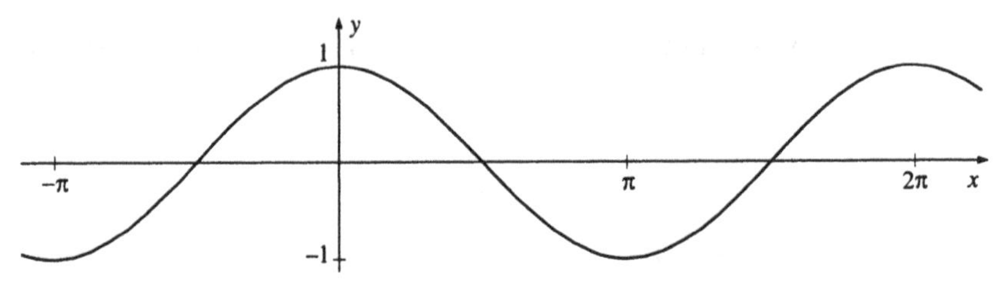

But $\sin(x + \pi/2) = \cos x$. The reader is invited to check this, either by using the addition formulas or by looking at the definitions, quadrant by quadrant. That is:

> *The graph of the function $y = \cos x$ can be obtained from the graph $y = \sin x$ by a shift to the left of $\pi/2$.*

We don't need to make a separate study of the curves $y = a\cos k(x+\beta)$. Letting $\gamma = x+\beta+\pi/2$, we can write any such curve as $y = a\sin k(x+\gamma)$.

For $n = 2$, we are graphing $y = \sin(x + \pi)$:

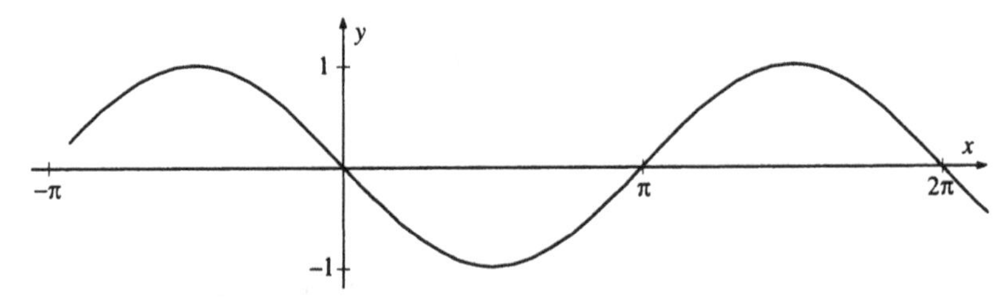

But $\sin(x+\pi) = -\sin x$. So we have:

> *The graph of the function $y = -\sin x$ can be obtained from the graph $y = \sin x$ by a shift to the left of π.*

In fact, we do not need to make a separate study of the curves $y = a\sin k(x+\gamma)$ for negative values of a. We need only adjust the value of γ, and we can describe each such curve with an equation in which $a > 0$.

The following general definition is convenient:

The number p is called a half-period of the function f if $f(x+p) = -f(x)$, for all values of x for which $f(x)$ and $f(x+p)$ are defined.

We have shown that π is a half-period of the function $y = \sin x$.

Now let $k = 3$. We obtain the following graph:

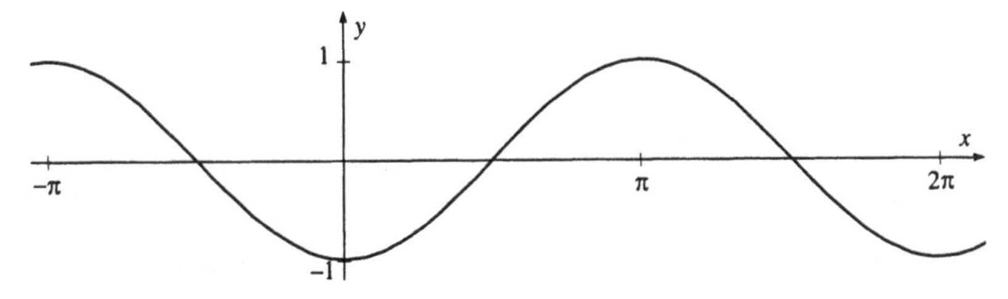

It is not difficult to check that

$$\sin(x + 3\pi/2) = -\cos x .$$

If $k = 4$, we will shift by $4(\pi/2) = 2\pi$, which we already know is a full period, and we will have come back to our original sine graph.

What if $k = 5$? Since $5 = 1 + 4$, we have $\sin(x + 5\pi/2) = \sin(x + \pi/2 + 4\pi/2) = \sin(x + \pi/2)$, because $2p$ is a period of the sine function. So $k = 5$ has the same effect as $k = 1$, and the cycle continues.

In general, we can make the following statements:

If $k = 4n$ for some integer n, then $\sin(x + k\pi/2) = \sin x$.

If $k = 4n + 1$ for some integer n, then $\sin(x + k\pi/2) = \cos x$.

If $k = 4n+2$ for some integer n, then $\sin(x+k\pi/2) = -\sin x$.

If $k = 4n+3$ for some integer n, then $\sin(x+k\pi/2) = -\cos x$.

To summarize, we have now examined the whole family of sinusoidal curves $y = a \sin k(x - \beta)$.

The constant a is called the *amplitude* of the curve. It tells us how far from 0 the values of the function can get. Without loss of generality, we may take a to be positive.

The constant k is called the *frequency* of the curve. It tells us how many periods are repeated in an interval of 2π. The period of the curve is $2\pi/k$. Without loss of generality, we can take k to be positive.

The constant β is called the *phase* or *phase shift* of the curve. It tells us how much the curve has been shifted right or left. If we allow β to be arbitrary, we need not consider negative values of a or k, and we need not study separately curves expressed using the cosine function.

Exercises

1–10: These exercises are multiple choice. Choose the answer

(A) if the given expression is equal to $\sin x$,

(B) if the given expression is equal to $\cos x$,

(C) if the given expression is equal to $-\sin x$, or

(D) if the given expression is equal to $-\cos x$.

1. $\sin(x + 2\pi)$
2. $\sin(x + 3\pi)$
3. $\sin(x + 9\pi/2)$
4. $\sin(x - \pi/2)$
5. $\sin(x - 3\pi/2)$
6. $\sin(x + 19\pi/2)$
7. $-\sin(x - 19\pi/2)$
8. $\sin(x + 157\pi/2)$
9. $\sin(x - 157\pi/2)$

11. Prove that π is a half-period of the function $y = \cos x$. Is π a half-period of the function $y = \tan x$? of $y = \cot x$?

12. Prove that if q is a half-period of some function f, then $2q$ is a period of f.

13. Show that for all values of x, $\cos(x + k\pi/2) =$

a) $-\sin x$, if $k = 4n + 1$ for some integer n,

b) $-\cos x$, if $k = 4n + 2$ for some integer n,

c) $\sin x$, if $k = 4n + 3$ for some integer n,

d) $\cos x$, if $x = 4n$ for some integer n.

14. Write each of the following in the form $y = a \sin k(x - \beta)$, where a and k are nonnegative:

a) $y = -2 \sin x$

b) $y = -2 \sin(x - \pi/3)$

c) $y = -2 \sin(x + \pi/4)$

d) $y = 3 \cos x$

e) $y = 3 \cos(x - \pi/6)$

f) $y = -3 \cos(x + \pi/8)$

15. Draw the graph of the function $y = \cos(x - \pi/5)$

16. Suppose we start with the graph of the function $y = \cos x$. By how much must we shift this graph to the right in order to obtain the graph of $y = \sin x$? By how much must we shift to the left to obtain the graph of $y = \sin x$?

17. Show that if k is odd, $\tan(x + k\pi/2) = -\cot x$. How can we simplify the expression $\tan(x + k\pi/2)$ if k is even?

8 Graphing the tangent and cotangent functions

The function $y = \tan x$ is different from the functions $y = \sin x$ and $y = \cos x$ in two significant ways. First, the domain of definition of the sine and cosine functions is all real numbers. However, $\tan x$ is not defined for $x = n\pi/2$, where n is an odd integer.

Second, the sine and cosine functions are *bounded*: the values they take on are always between -1 and 1 (inclusive). But the function $y = \tan x$ takes on all real numbers as values.

These differences are easily seen in the graph of the function $y = \tan x$:

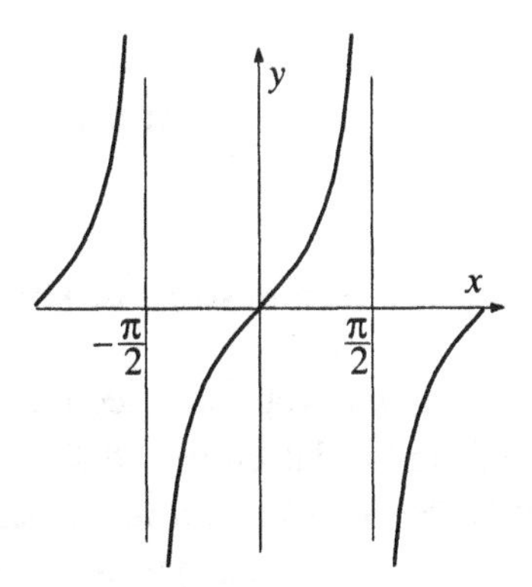

Note that the graph approaches the line $x = \pi/2$, but never reaches it. This line is called a *vertical asymptote* of the curve $y = \tan x$. This graph $y = \tan x$ has a vertical asymptote at every line $y = n\pi/2$, for n an odd integer.

To draw the graph of $y = \cot x$, we note that

$$\cot x = \frac{\cos x}{\sin x} = -\frac{\sin(x - \pi/2)}{\cos(x - \pi/2)} = -\tan(x - \pi/2).$$

Therefore, the graph of $y = \cot x$ also takes on all real numbers as values. It is not defined for $x = n\pi$, where n is any integer, and has vertical asymptotes at $y = n\pi$:

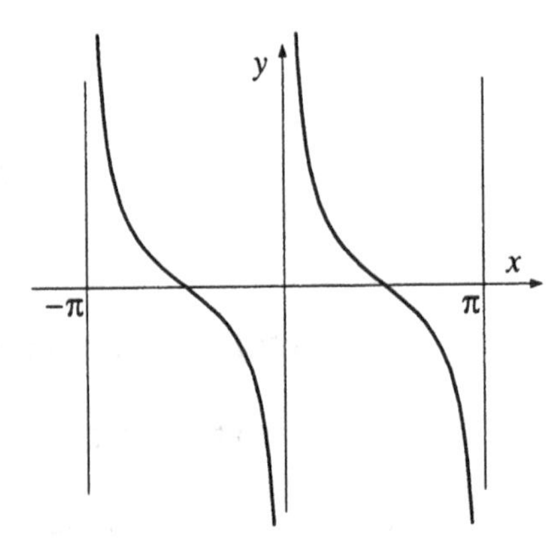

Exercises

1. Draw the graphs of

   a) $y = \tan(x - \pi/6)$ b) $y = 3\tan x$ c) $y = \cot(x + \pi/4)$.

2. Suppose we graphed the equation $y = \tan x$. Is it possible to describe this graph with an equation of the form $y = \cot(x + \varphi)$, for some number φ? Why or why not?

9 An important question about sums of sinusoidal functions

We hope that from this material you have seen the importance, and the beauty, of the family of sinusoidal curves that we have been studying. Physicists call this family the curves of *harmonic oscillation*.

Let us now consider the following question. Suppose we have two sinusoidal curves (harmonic oscillations):

$$y_1 = a_1 \sin k_1(x - \beta_1)$$
$$y_2 = a_2 \sin k_2(x - \beta_2).$$

Will the sum of these two also be a sinusoidal curve (harmonic oscillation)? That is, will

$$y = a_1 \sin k_1(x - \beta_1) + a_2 \sin k_2(x - \beta_2)$$

be a sinusoidal curve? The answer is somewhat surprising. If $k_1 = k_2$, the answer is yes, but if $k_1 \neq k_2$, the answer is no.

That is, the sum of two harmonic oscillations is again a harmonic oscillation if and only if the original frequencies are the same. The results of the next few sections will allow us to explore this situation.

Exercises

Each of these exercises concerns the following three functions:

$$y_1 = 2\sin x$$
$$y_2 = \sin(x - \pi/4)$$
$$y_3 = 3\sin 2x$$

1. Use your calculator to draw the graph of (a) $y_1 + y_2$; (b) $y_1 + y_3$; (c) $y_2 + y_3$.
2. Which of the graphs in Exercise 1 appear to be sinusoidal functions?

10 Linear combinations of sines and cosines

Definition: If we have two functions $f(x)$ and $g(x)$, and two constants a and b, then the expression $af(x) + bg(x)$ is called a *linear combination* of the functions $f(x)$ and $g(x)$.

Let us look at the graph of a linear combination of sinusoidal curves.

Example 65 Graph the function $y = \frac{\sqrt{3}}{2}\sin x + \frac{1}{2}\cos x$.

Solution. Since $\frac{\sqrt{3}}{2} = \cos\frac{\pi}{6}$ and $\frac{1}{2} = \sin\frac{\pi}{6}$, we use the formula $\sin(\alpha + \beta) = \sin\alpha\cos\beta + \cos\alpha\sin\beta$. Letting $\alpha = x$ and $\beta = \frac{\pi}{6}$, this formula tells us that the given function can be written as $y = \sin(x + \frac{\pi}{6})$. Now we can graph it as we did in Section 5:

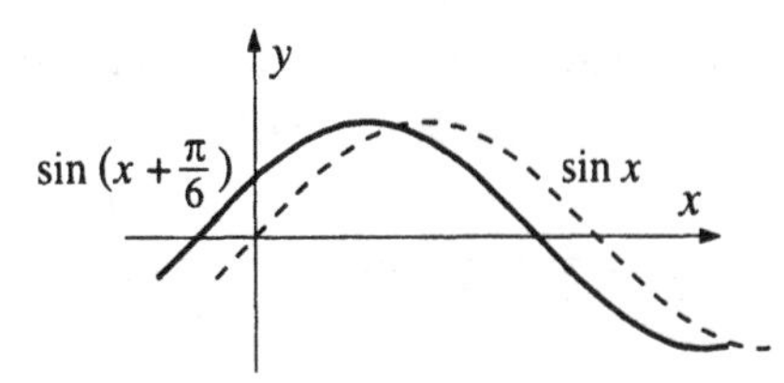

This solution may seem artificial, but is in fact a general method. It works because there is an angle φ such that $\cos\varphi = 1/2$ and $\sin\varphi = \sqrt{3}/2$, and this happened because the values $A = 1/2$ and $B = \sqrt{3}/2$ satisfy the equation $A^2 + B^2 = 1$. (The reader is invited to do this computation.)

But what if $A^2 + B^2$ is not equal to 1?

Example 66 Draw the graph of the function $f(x) = 3\sin x + 4\cos x$.

Solution. Our "best friends" (of Chapter 1) are hiding in this expression: where we have 3 and 4, we try to look for the number 5. Indeed, $f(x)/5 = \frac{3}{5}\sin x + \frac{4}{5}\cos x$, and $\left(\frac{3}{5}\right)^2 + \left(\frac{4}{5}\right)^2 = 1$, so we can use the method of Example 3. We know that there is an angle φ such that $\cos\varphi = 3/5$ and $\sin\varphi = 4/5$, and so

$$\frac{f(x)}{5} = \cos\varphi \sin x + \sin\varphi \cos x = \sin(x + \varphi)$$

or $f(x) = 5\sin(x + \varphi)$, for a certain angle φ. The graph is a sine curve, shifted to the left φ units, and with amplitude 5:

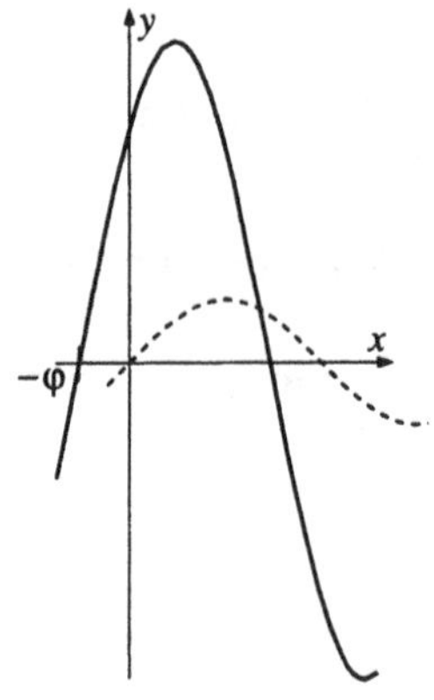

The same technique will work for linear combinations of $y = \sin kx$ and $y = \cos kx$, as long as the frequency of the two functions is the same. This is important enough to state as a theorem:

Theorem A linear combination of $y = \sin kx$ and $y = \cos kx$ can be expressed as $y = a\sin k(x + \varphi)$, for suitable constants a and φ.

Proof. A linear combination of $y = \sin kx$ and $y = \cos kx$ has the form $y = A\sin kx + B\cos kx$. We can rewrite it as

$$y = \sqrt{A^2 + B^2}\left(\frac{A}{\sqrt{A^2 + B^2}}\sin kx + \frac{B}{\sqrt{A^2 + B^2}}\cos kx\right).$$

Then

$$\left(\frac{A}{\sqrt{A^2 + B^2}}\right)^2 + \left(\frac{B}{\sqrt{A^2 + B^2}}\right)^2 = 1$$

so there exists an angle α such that

$$\cos\alpha = \frac{A}{\sqrt{A^2+B^2}} \text{ and } \sin\alpha = \frac{B}{\sqrt{A^2+B^2}} . \tag{1}$$

Now we can write

$$\begin{aligned} A\sin kx + B\cos kx &= \sqrt{A^2+B^2}(\cos\alpha \sin kx + \sin\alpha \cos kx) \\ &= \sqrt{A^2+B^2}\sin(kx+\alpha) \\ &= \sqrt{A^2+B^2}\sin k(x+\alpha/k) . \end{aligned}$$

Taking $a = \sqrt{A^2+B^2}$ and $\varphi = \alpha/k$, we have the required form. □

We have proved that $A\sin kx + B\cos kx$ can be written in the form $a\sin k(x+\gamma)$, where $a = \sqrt{A^2+B^2}$ and $\varphi = \alpha/k$ (for α defined by equations (1) above).

The converse statement is also correct:

Theorem The function $a\sin k(x+\varphi)$ can be written as a linear combination of the functions $\sin kx$ and $\cos kx$.

Proof. We have $a\sin k(x+\varphi) = a(\sin kx\cos k\varphi + \cos kx \sin k\varphi)$. Taking $A = a\cos k\varphi$ and $B = a\sin k\varphi$, we see that $a\sin k(x+\varphi) = A\sin kx + B\cos kx$. □

We can now write a sinusoidal curve in either of two standard forms: $y = a\sin k(x-\beta)$ or $y = A\sin kx + B\cos kx$.

Example 67 Write the function $y = 2\sin(x+\pi/3)$ as a linear combination of the function $y = \sin x$ and $y = \cos x$.

Solution. We have $2\sin(x+\pi/3) = 2(\sin x\cos\pi/3 + \cos x\sin\pi/3) = 2(1/2)\sin x + 2\sqrt{3}/2\cos x = \sin x + \sqrt{3}\cos x$.

Exercises

1. Write the function $y = 2\sin x + 3\cos x$ in the form $y = a\sin k(x-\beta)$. What is its amplitude?

2. What is the maximum value achieved by the function $y = 2\sin x + 3\cos x$?

3–6: Write each function in the form $y = a \sin k(x - \beta)$. What is the maximum value of each function?

3. $y = \sin x + \cos x$
4. $y = \sin x - \cos x$
5. $y = 4 \sin x + 3 \cos x$
6. $y = \sin 2x + 3 \cos 2x$

7, 8: Write each function in the form $A \sin x + B \cos x$:

7. $y = \sin(x - \pi/4)$
8. $y = 4 \sin(x + \pi/6)$

11 Linear combinations of sinusoidal curves with the same frequency

Now we are ready to address the important question of Section 9.

Theorem The sum of two sinusoidal curves with the same frequency is again a sinusoidal curve with this same frequency.

Proof. Let us take the two sinusoidal curves

$$a_1 \sin k(x - \beta_1) \text{ and}$$
$$a_2 \sin k(x - \beta_2) .$$

Using the addition formula, we can write:

$$a_1 \sin k(x - \beta_1) = A_1 \sin kx + B_1 \cos kx$$
$$a_2 \sin k(x - \beta_2) = A_2 \sin kx + B_2 \cos kx$$

for suitable values of A_1, A_2, B_1, and B_2. Then our sum is equal to

$$(A_1 + A_2) \sin kx + (B_1 + B_2) \cos kx .$$

But we know, from the theorem of Section 9, that this sum is also a sinusoidal curve. Our theorem is proved. □

We invite the reader to fill in the details, by giving the expressions for A_1, A_2, B_1, and B_2.

Note that the two functions we are adding may have different amplitudes. The result depends only on their having the same period. This result is very important in working with electricity. Alternating electric current is described by a sinusoidal curve, and this theorem says that if we add two currents with the same periods, the resulting current will have this period as well. So if we are drawing electric power from different sources, we need not worry how to mix them (whether their phase shifts are aligned), as long as their periods are the same.

The next result is important in more advanced work:

Theorem If a linear combination of the functions $y = \sin kx$ and $y = \cos kx$ is shifted by an angle β, then the result can be expressed as a linear combination of the same two functions.

Proof. Let us take the linear combination

$$a \sin kx + b \cos kx$$

and shift it by an angle β. The result is

$$a \sin k(x - \beta) + b \cos k(x - \beta) .$$

We know that $\cos k(x-\beta)$ can be written as $\sin k(x-\gamma)$, for some angle γ. Thus we can write our shifted linear combination as

$$a \sin k(x - \beta) + b \sin k(x - \gamma) .$$

But this is a sum of sinusoidal curves with the same frequency k, so the previous theorem tells us that it can be written as a single sinusoidal curve with frequency k (even though the shifts are different!). And we know, from Section 9, that such a sum can be written as a linear combination of $\sin kx$ and $\cos kx$.

Example 68 Suppose we take the graph of a linear combination of $y = \sin x$ and $y = \cos x$:

$$y = 2 \sin x + 4 \cos x$$

and shift it $\pi/6$ units to the left. We get:

$$\begin{aligned} y &= 2 \sin(x + \pi/6) + 4 \cos(x + \pi/6) \\ &= 2(\sin x \cos \pi/6 + \cos x \sin \pi/6) + 4(\cos x \cos \pi/6 - \sin x \sin \pi/6) \\ &= 2(\sqrt{3}/2 \sin x + 2(1/2) \cos x + 4(\sqrt{3}/2) \cos x - 4(1/2) \sin x \\ &= (\sqrt{3} - 2) \sin x + (2\sqrt{3} + 1) \cos x \end{aligned}$$

which is again a linear combination of $y = \sin x$ and $y = \cos x$.

This technique works whenever we apply a shift to a linear combination of $y = \sin kx$ and $y = \cos kx$. The proof follows the reasoning of the above example.

A final comment: We have not considered linear combinations of sines and cosines with *different* frequencies. This is a more difficult situation, and leads to some very advanced mathematical topics, such as Fourier Series and almost periodic functions. We will return to this question a bit later.

Exercises

1. Express each function in the form $y = A \sin kx + B \cos kx$

   (a) $y = 2\sin(x + \pi/6) + \cos(x + \pi/6)$

   (b) $y = 2\sin 2(x + \pi/4) - \cos 2(x + \pi/4)$

2. Look at the exercises for Section 9 on page 189.

   (a) Write $y_1 + y_2$ as a linear combination of $\sin x$ and $\cos x$.

   (b) What goes wrong when you try to write $y_1 + y_3$ as a linear combination of $\sin x$ and $\cos x$?

12 Linear combinations of functions with different frequencies

So far, we have some important results about linear combinations of sines and cosines with the same frequency. We would like to investigate the sum of two functions like $y = \sin k_1 x$ and $y = \sin k_2$, where $k_1 \neq k_2$. We start the discussion with some examples which may not at first appear related.

Example 69 Graph the function $y = x + \sin x$.

Solution. Each y-value on this graph is the sum of two other y-values: the value $y = \sin x$ and the value $y = x$. So we can take each point on the curve $y = \sin x$ and "lift it up" by adding the value $y = x$ to the value $y = \sin x$.

This is particularly easy to see for those points where $\sin x = 0$. For these points, the value of $x + \sin x$ is just x:

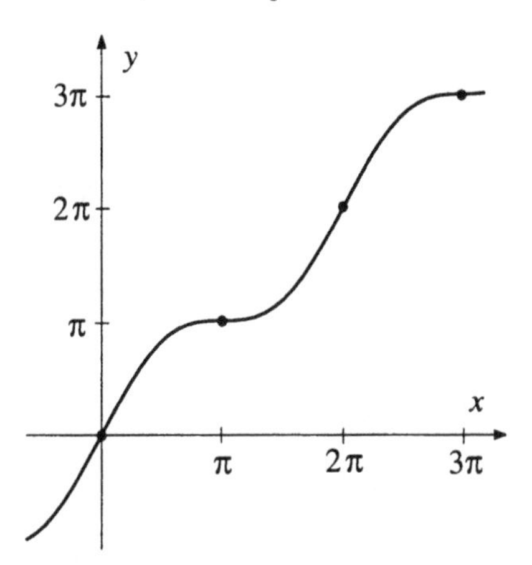

In between these points, the line $y = x$ is lifted up slightly, or brought down slightly, by positive or negative values of $\sin x$. We can think of the sine curve as "riding" on the line $y = x$.

Example 70 Graph the function $y = \sin x + 1/10 \sin 20x$.

Solution. This seems much more complicated, but in fact can be solved using the same method as the previous examples. We graph the two curves $y = \sin x$ and $y = 1/10 \sin 20x$ independently, then add their y-values at each point:

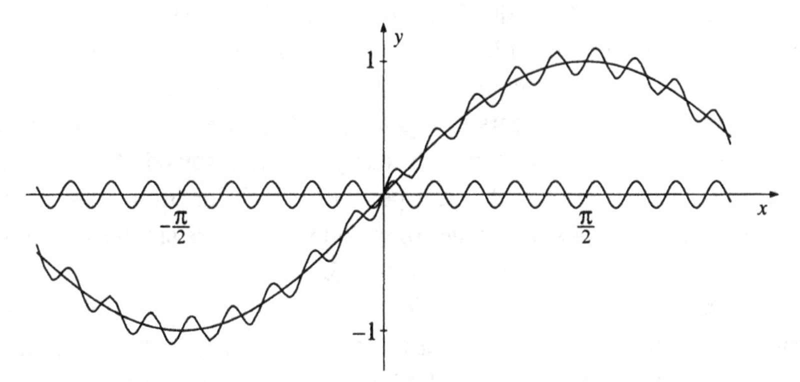

Again, we can think of one curve "riding" on the other. This time the curve $y = 1/10 \sin 20x$ "rides" on the curve $y = \sin x$, or *perturbs* it a bit at each point.

Note that our new curve is *not* a sinusoidal curve. We cannot express it either in the form $y = a \sin k(x - \beta)$ or in the form $y = A \sin kx + b \cos kx$.

Exercises

Construct graphs of the following functions.

1. $y = -x + \sin x$
2. $y = x^2 + \sin x$
3. $y = x^2 + \cos x$. Hint: Is the function odd? Is it even?
4. $y = x^3 + \sin x$
5. $y = x^2 + (1/10) \sin x$
6. $y = \cos x + (1/10) \sin 20x$
7. $y = 2 \sin x + (1/10) \sin 20x$

13 Finding the period of a sum of sinusoidal curves with different periods

We know that the function $y = \sin 10x + \sin 15x$ is not a sinusoidal curve. Let us show that it is still periodic. Indeed, if we shift the curve by 2π, we have $y = \sin 10(x + 2\pi) + \sin 15(x + 2\pi) = \sin(10x + 20\pi) + \sin(15x + 30\pi) = \sin 10x + \sin 15x$.

But what is its smallest positive period? We can answer this by looking separately at all the periods of the two functions we are adding. Any period of $y = \sin 10x$ must have the form $m(2\pi/10)$, for some integer m. Any period of $y = \sin 15x$ must have the form $n(2\pi/15)$, for some integer n. To be a period of both functions, a number must be of both these forms. That is, we must have integers m and n such that $2m\pi/10 = 2n\pi/15$, or $3m = 2n$. If we take $m = 2$, $n = 3$, our problem is solved. The number $2\pi/5 = 2(2\pi/10) = 3(2\pi/15)$ is a period for both functions. And since we took the smallest positive values of m and n, this is the smallest positive period for the function $y = \sin 10x + \sin 15x$.

The argument above is drawn from number theory, where it is connected with the least common multiple of two numbers. This concept is used in elementary arithmetic, in finding the least common denominator for two fractions. The general statement, proved in number theory, is this:

> *The function $y = \sin k_1 x + \sin k_2 x$ is periodic if and only if the quotient k_1/k_2 is rational.*

But a function like $y = \sin x + \sin \sqrt{2}x$ has no period at all.

Exercises

Find the (smallest positive) period for each of the following functions.

1. $y = \sin 2x + \sin 3x$
2. $y = \sin 3x + \sin 6x$
3. $y = \sin 4x + \sin 6x$
4. $y = \sin \sqrt{2}x + \sin 3\sqrt{2}x$

14 A discovery of Monsieur Fourier

Example 71 Graph the function $y = \sin x + (1/3) \sin 3x$.

Solution. This example is similar to Example 70. The values of $\sin x$ are "perturbed" by those of $\frac{1}{3} \sin 3x$:

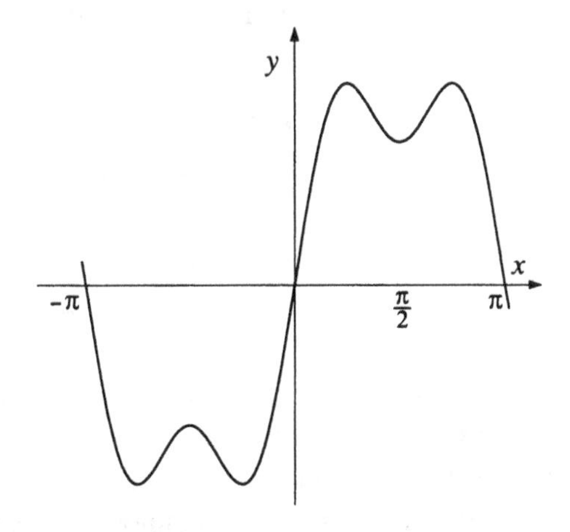

Example 72 Graph the function $y = \sin x + \frac{1}{3} \sin 3x + \frac{1}{5} \sin 5x$. (Use a graphing calculator or software utility for this complicated function.)

Solution.

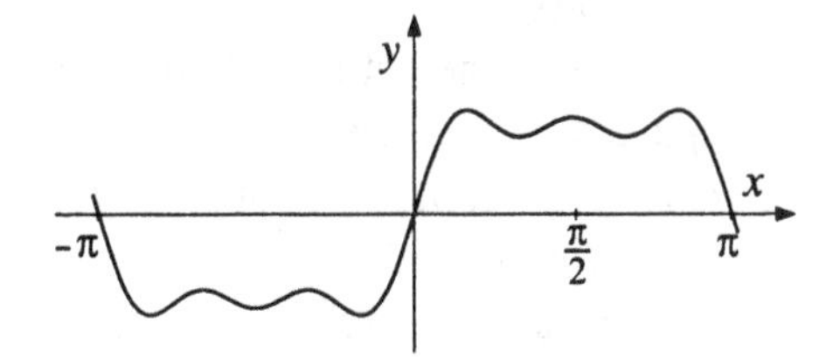

Let us compare the graphs of the three functions:

$$y = \sin x$$
$$y = \sin x + \tfrac{1}{3}\sin 3x$$
$$y = \sin x + \tfrac{1}{3}\sin 3x + \tfrac{1}{5}\sin 5x\ .$$

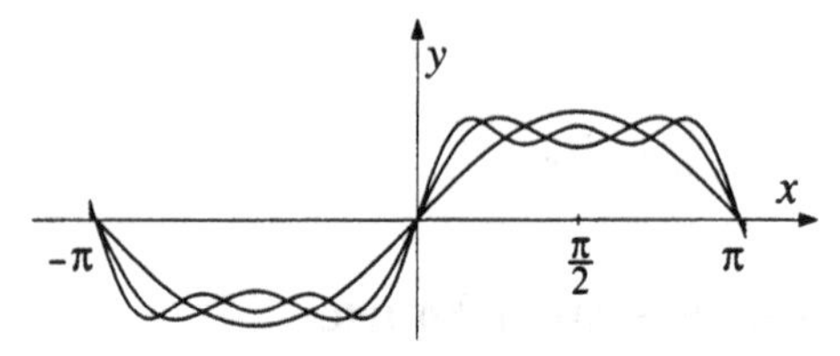

The formulas for these functions show a pattern. Can you guess what the next formula in the pattern would be? Can you guess what its graph would look like? Check your guess with a graphing calculator or software utility.

It is not difficult to guess that the graphs of these functions will look more and more like the following:

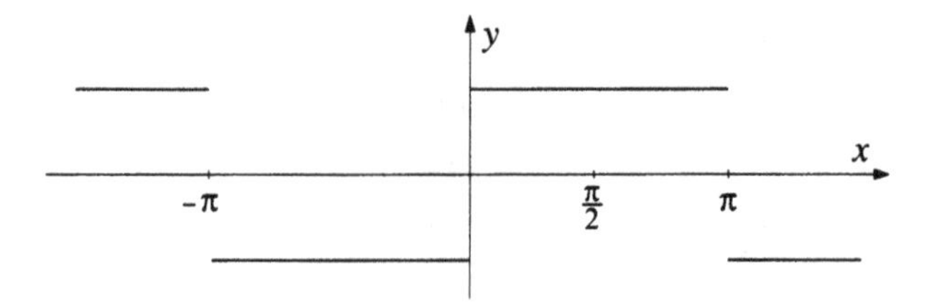

Mathematicians say that this sequence of functions *converges* to a limit, and that this limit is the function whose graph is given above. In fact, this is a special case of the very important mathematical theory of Fourier series. The French physicist Fourier discovered that almost any periodic function, including some with very complicated or bizarre graphs, can be represented as the limit of a sum of sines and cosines (the above example doesn't happen to contain cosines). He also showed how to calculate this sum (using techniques drawn from calculus).

Fourier's discovery allows mathematicians to describe very simply any periodic function, and physicists can use these descriptions to model actions that repeat. For example, sounds are caused by periodic vibrations of particles of air. Heartbeats are periodic motions of a muscle in the body. These phenomena, and more, can be explored using the mathematical tools of Fourier analysis.

Exercises

Please use a graphing calculator or graphing software package for these exercises.

1. Graph the function $y = \sin x - \frac{1}{2}\sin 2x + \frac{1}{3}\sin 3x$.

2. Graph the function $y = \sin x - \frac{1}{2}\sin 2x + \frac{1}{3}\sin 3x - \frac{1}{4}\sin 4x + \frac{1}{5}\sin 5x$.

3. Consider the sequence of functions:

$$y = \sin x$$
$$y = \sin x - \tfrac{1}{2}\sin 2x$$
$$y = \sin x - \tfrac{1}{2}\sin 2x + \tfrac{1}{3}\sin 3x$$
$$y = \sin x - \tfrac{1}{2}\sin 2x + \tfrac{1}{3}\sin 3x - \tfrac{1}{4}\sin 4x$$
$$y = \sin x - \tfrac{1}{2}\sin 2x + \tfrac{1}{3}\sin 3x - \tfrac{1}{4}\sin 4x + \tfrac{1}{5}\sin 5x$$

   Draw the graph of the function that you think is the limit of this sequence of functions.

4. Consider the sequence of functions:

$$y = \cos x$$
$$y = \cos x + \tfrac{1}{9}\cos 3x$$
$$y = \cos x + \tfrac{1}{9}\cos 3x + \tfrac{1}{25}\cos 5x$$
$$y = \cos x + \tfrac{1}{9}\cos 3x + \tfrac{1}{25}\cos 5x + \tfrac{1}{49}\cos 7x$$

   Draw the graph of the function that you think is the limit of this sequence of functions. Do you recognize the pattern in the amplitudes?

Appendix

I. Periodic phenomena

Many phenomena in nature exhibit periodic behavior: the motions repeat themselves after a certain amount of time has passed. The sine function, it turns out, is the key to describing such phenomena mathematically.

The following exercises concern certain periodic motions. Their mathematical representations remind us of the sine curve, but are not exactly the same. In more advanced work trigonometric functions can indeed be used to describe these motions.

Exercises

1. The diagram represents a line segment 1 foot in length.

An ant is walking along the line segment, from point A to point B, then back again, at a speed of one foot per minute. Draw a graph showing the distance from point A to the ant's position at a given time t. For example, when $t = 0.5$, the ant is halfway between A and B, and headed towards B.

2. The diagram shows part of a number line, from $A = -1$ to $P = 0$, to $B = +1$.

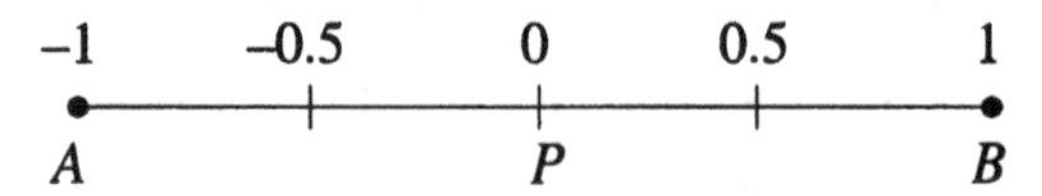

An ant is walking along the number line, starting from point P. The ant walks to point B, then to A, then back to B, and so on. The ant walks at a speed of one foot per minute. Draw a graph showing the position of the ant on the number line at time t. For example, when $t = 0.5$, the ant is at 0.5, and when $t = 2.5$, the ant is at -0.5.

3. The diagram shows a square wall of a room. Each side of the square is 8 feet long.

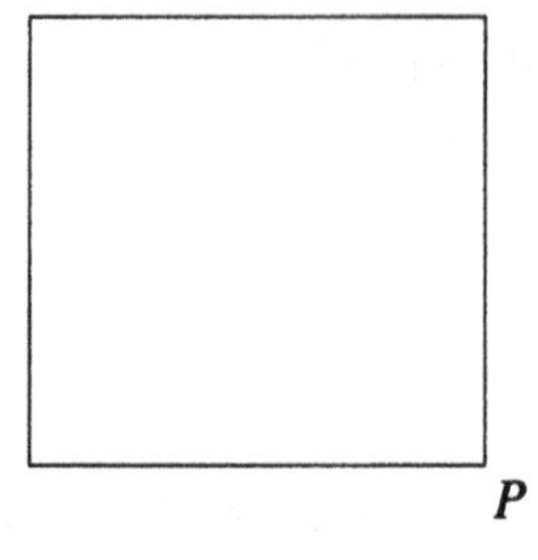

An ant is walking along the perimeter of the wall, at a speed of one foot per minute, starting at the point P shown and moving counterclockwise. Draw a graph showing the height of the ant above the floor (call it h) at any given time t. For example, where $t = 4$ the ant's height is 4 feet, and where $t = 12$ the ant's height is 8 feet.

4. The Bay of Fundy lies between the Canadian provinces of Nova Scotia and New Brunswick. The people who live on its shores experience some of the world's highest tides, which can reach a height of 40 feet. This creates a landscape that shifts twice every day. Beaches become bays, small streams turn into raging rivers, and peninsulas are suddenly islands as the tides rise and fall.

   For anyone who lives near the ocean, it is important to know when high and low tide will occur. But for the Bay of Fundy, it is critical also to know how fast the tide is rising or falling. The inhabitants of this area use the so-called *rule of twelfths* to estimate this. They take the interval between low and high tide to be 6 hours (it is actually a bit more). Then they approximate that:

   $\frac{1}{12}$ of the tide will come in during the first hour
   $\frac{2}{12}$ of the tide will come in during the second hour
   $\frac{3}{12}$ of the tide will come in during the third hour
   $\frac{3}{12}$ of the tide will come in during the fourth hour
   $\frac{2}{12}$ of the tide will come in during the fifth hour
   $\frac{1}{12}$ of the tide will come in during the sixth hour

   Assume that the height of a day's tide is 36 feet, and draw a graph of the height of the water at a given point along the Bay of Fundy, using these estimates.

   When is the tide running fastest? Slowest?

II. How to explain the shifting of the graph to your younger brother or sister

When we were little, we used to go every few months to the doctor. The doctor would measure our height, and make a graph showing how tall we were at every visit. Here is the graph for my height :[2]

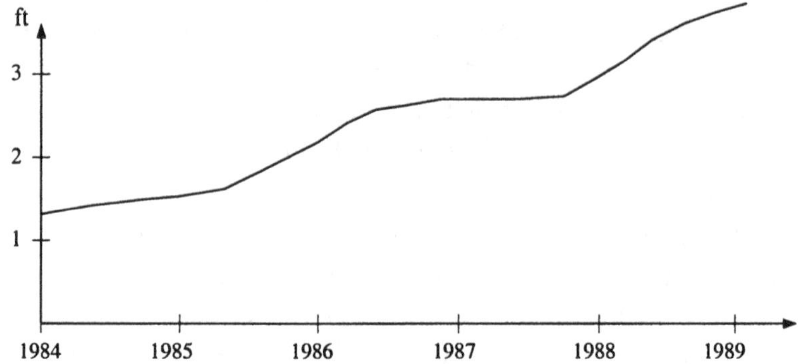

Two years later, when you were born, our parents asked the doctor if she could predict your growth year-by-year. Well, she couldn't exactly do this, but she said: "If the new baby follows the same growth curve as your older child, then he will be as tall as the older one was three years earlier." So the doctor was predicting a growth curve for you which looks like this:

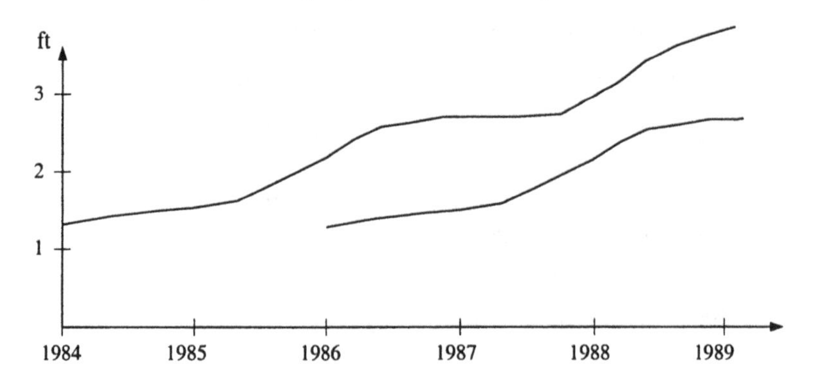

You will be 3 feet tall exactly two years after I was 3 feet tall, and 4 feet tall also exactly two years after I was, and so on. Your graph is the same as mine, but shifted to the right by two years. If you want to know the prediction for your height, just look at what my height was two years ago. So if my graph is described by the equation height $= f(\text{year})$, then your graph is described by the equation height $= f(\text{year} - 2)$.

Of course, it hasn't quite turned out this way. My growth curve was not exactly the same as yours. So the doctor's prediction was not accurate. But for some families, it is accurate.

---

[2]Of course, when I was very little, I couldn't stand up, so they measured my "length." When I learned to stand, this became my height.

In just the same way, if you are dealing with the graph $y = \sin(x - \alpha)$, rather than $y = \sin x$, you must "wait" for x to get bigger by α before the height of the new graph is the same as that of the old graph. So the new graph is shifted α units to the right.

III. Sinusoidal curves with rational periods

We have taken, as our basic sine curve, the function $y = \sin x$. The period of this function is 2π, which is an irrational number. The other functions we've investigated also have irrational periods. Can a sine curve have a rational period?

Consider the function $y = \sin 2\pi x$. Using our formula, its period is $2\pi/2\pi = 1$. We can check this directly:

$$\sin\big(2\pi(x+1)\big) = \sin(2\pi x + 2\pi) = \sin 2\pi x .$$

The exercises below require the construction of sinusoidal curves with other rational periods.

Exercises

1. Show that the function $y = \sin \pi x$ has the value 0 when $x = 1$, $x = 2$, $x = 3$, and $x = 4$.

2. Show that the function $y = \sin 4\pi x$ has a period of $\frac{1}{2}$.

3. Write the equation of a sine curve with period 3.

4. Write the equation of a sine curve with period 2.

5. If n is a positive integer, write a function of the form $y = \sin kx$ with period n.

IV. From graphs to equations

A tale is told of the Russian tsar Alexei Mikhaelovitch, the second of the Romanov line (1629–76; reigned 1645–76). His court astronomer came to him one day in December, and told him, "Your majesty, from this day forth the number of hours of daylight will be increasing."

The tsar was pleased. "You have done well, court astronomer. Please accept this gift for your services." And, motioning to a courtier, he presented the astronomer with a valuable gemstone.

The astronomer enjoyed his gift and practiced his arts, until one day in June, when he again reported to the tsar. "Your highness, from this day forth the number of hours of daylight will be be decreasing."

The tsar scowled. "What? More darkness in my realm?" And he ordered the hapless astronomer beaten.

Of course, the variation in the amount of daylight was not the fault of this astronomer, or any other astronomer. It is due to the circumstance that the earth's axis is tilted with respect to the plane in which it orbits the sun. Because of this phenomenon, the days grow longer from December to June, then shorter from June to December.

What is interesting to us is the rate at which the number of hours of daylight changes. It turns out that if we graph the number of hours of daylight in each day, we get a sinusoidal curve:

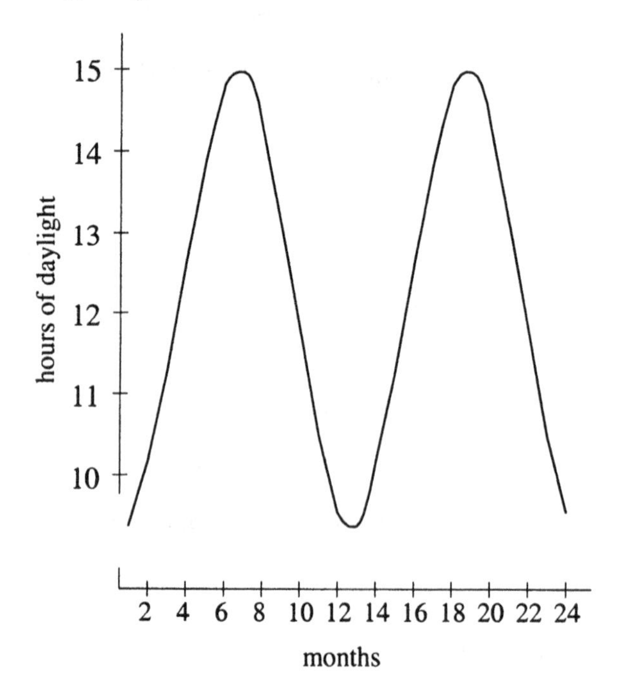

Since this curve is high above the x-axis, we have shown it with the y-axis "broken," so that you can see the interesting part of the graph. If you don't like this, try redrawing the curve without a "broken" y-axis. You will find that most of your diagram is empty.

We will learn more about this curve in the following exercises.

Exercises

1. By estimating the distances on the graph above, find an equation of the form $y = a \sin k(x - \pi)$ which approximates the function whose graph is shown.

2. The curves below give the hours of daylight at certain latitudes.

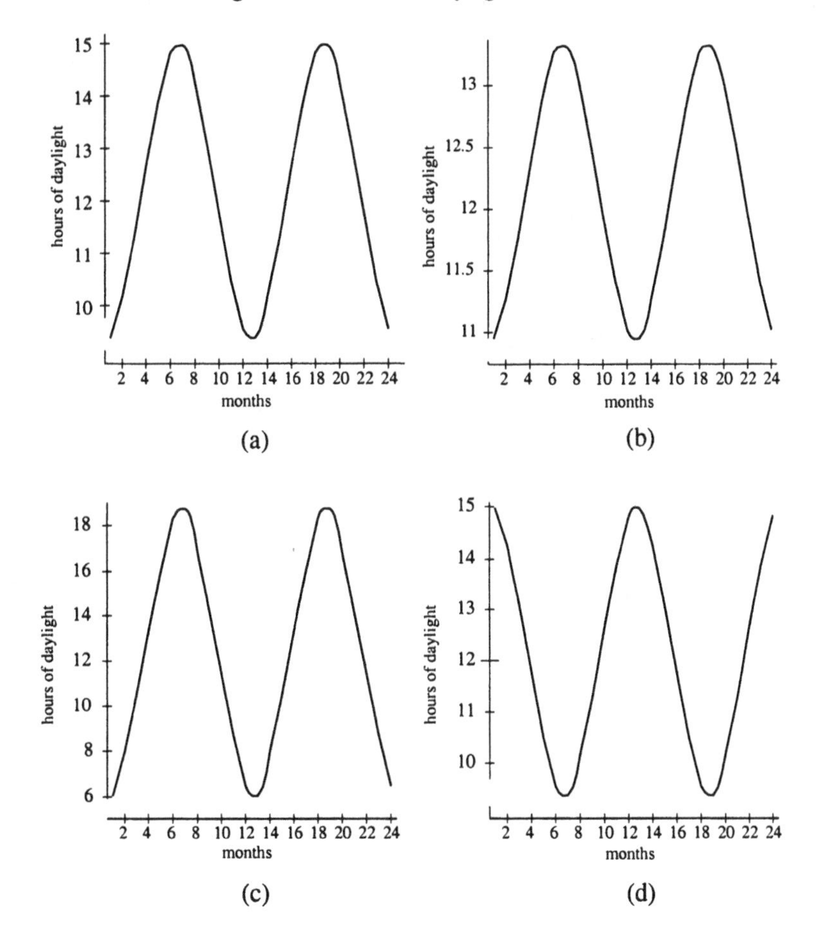

Notice that the maximum number of hours of daylight occur at the same time of year from graphs (a), (b), and (c), but at different times for graph (d). If graph (a) corresponds to a location in the northern hemisphere, in which hemisphere are the locations of the other graphs?

3. Notice that the "average" number of daylight hours is the same for each graph. This "average" is given by the y-coordinates along the line around which the curve oscillates: On certain days of the year, at each location, the actual number of hours of daylight is the same as the average number. How does the time of year at which this average is actually achieved vary from location to location?

Chapter 9

Inverse Functions and Trigonometric Equations

1 Functions and Inverse Functions

Let us recall the definition of a function. If we have two sets A and B, a function from set A to set B is a correspondence between elements of A and elements of B such that

1. Each element of A corresponds to some element of B, and
2. No element of A corresponds to more than one element of B.

If the element x in set A corresponds to the element y in B, we write $y = f(x)$, where f is the symbol for the function itself.

Example 73 Let us take A as the set of all real numbers, and B as another copy of the set of real numbers. If x is an element of A, then we can make it correspond to an element y in B by taking $y = x^2$. Every element x in A corresponds to some element y in B, (since any number can be squared), and no element x in A corresponds to more than one element y in B (since we get a unique answer when we square a number).

Our definition of a function is not very democratic. For every element of A, we must produce exactly one element of B. But if we have an element of B, we cannot tell if there is an element in A to which it corresponds. An element of B may correspond to no element of A, to one element of A, or to more than one element of A.[1]

[1] In older texts, this undemocratic situation was described by calling x the *independent variable* and y the *dependent variable*.

Example 74 In Example 73, if we were given a number x in A, we are obliged to supply an answer to the question: what number y in B corresponds to A? For example, if $x = 3$, then we can answer that $y = 9$, and if $x = -3$, we can answer $y = 9$ again. This is allowed, under our definition of a function. The only restriction is that our answer must be a number in set B.

But if we choose an element y in set B, we are not obliged to answer the question: what number in A corresponds to it? Certainly, if we chose $y = 9$, we could answer $x = 3$. But we could just as well answer $x = -3$, and so our answer would not be unique. Worse, if we chose $y = -1$, we have no answer at all. There is no real number whose square is -1.

That is, if y is a function of x, it may not be the case that x is also a function of y. However, in some cases, we can improve the situation.

Example 75 Take the set A to be the set of nonnegative real numbers, and for B take another copy of the same set. As before, the correspondence $y = x^2$ is a function: if x is a number in A, then x^2 is a number in B, since the square of a real number cannot be negative. But now, if we take a number y in B, we can always answer the question: What number x in A corresponds to y? For example, if $y = 9$, we can answer that $x = 3$. We are not embarrassed by the possibility of a second answer, since -3 is not in our (new) set A. Nor are we embarrassed by the lack of any answer. Negative numbers, which are not squares of real numbers, do not exist in our new set B.

In general, we can take a function $y = f(x)$, try to start with a value of y, and get the corresponding value of x. If this is possible – if x is a function of y as well – then this new function is called the *inverse function* for $f(x)$.

Thus the function $y = x^2$, where $x \geq 0$ and $y \geq 0$, does have an inverse, given by the formula $x = \sqrt{y}$. This is the reason for insisting, in elementary algebra books, that the symbol $\sqrt{y}$ refers to the *nonnegative* real number whose square is y.

When does a function have an inverse function? This is an important question. We will not give a general answer here. We will, however, observe that if A and B are intervals on the real line, then $y = f(x)$, defined on these intervals, has an inverse if and only if it is *monotone* (steadily increasing or steadily decreasing). The first two graphs below show functions that are monotone, and have inverses. The last three graphs show functions that have no inverse on the sets A and B.

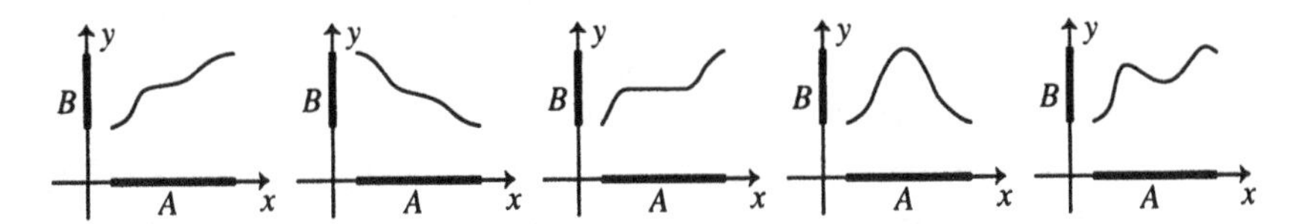

2 Arcsin: The inverse function to sin

Example 76 The equation $y = \sin x$ defines a function from the set A of real numbers to the set B of real numbers. Does it have an inverse function?

Again, the answer is no, and for the same two reasons as in Example 75. For some values of y in B, such as $y = 5$, there are no values of x such that $\sin x = y$. For other values of y, such as $y = 1/2$, there are many values of x: $\sin \pi/6 = 1/2$, $\sin 5\pi/6 = 1/2$, $\sin 13\pi/6 = 1/2$, and so on.

In Example 75, we were able to overcome these difficulties, by restricting the sets A and B that the function is defined on. Can we do this here? Let us look at the graph of $y = \sin x$.

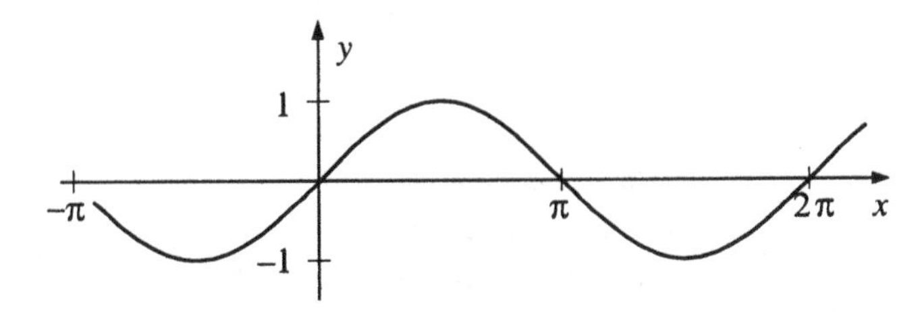

Let us start by including the number 0 in our set A. We must choose for set A a domain on which the function $y = \sin x$ is monotone, and it's easiest to take the for set A the set $-\pi/2 \leq x \leq \pi/2$:

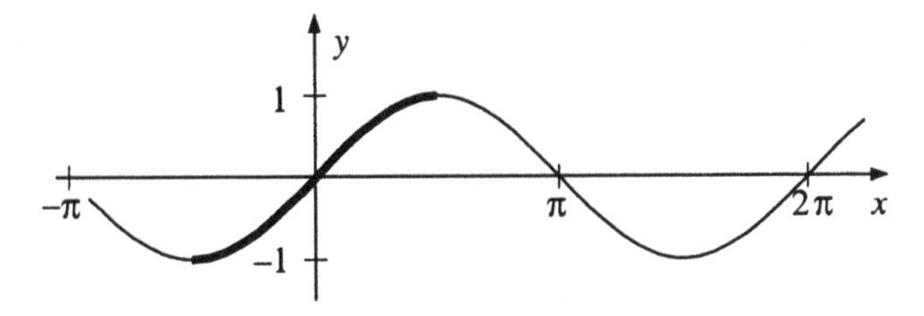

Now we can choose for set A our interval $-\pi/2 \leq x \leq \pi/2$, and for set B the interval $-1 \leq y \leq 1$, and for every y in set B, there exists exactly one x in A such that $\sin x = y$.

The inverse function to $y = \sin x$, defined in this way, is important enough to merit its own name. It is called the *arcsine* function[2], and if $y = \sin x$ (with x and y in the two sets described above), we write $x = \arcsin y$.

However, sometimes we will discuss the arcsine function in its own right. Then we will write $y = \arcsin x$, where $-1 \leq x \leq 1$, and $-\pi/2 \leq y \leq \pi/2$. We have already had a chance encounter with this function on the calculator. Now we will get to know it much better.

Example 77 Find $\arcsin 1/2$.

Solution: Again, if $y = \arcsin 1/2$, then $\sin y = 1/2$. There are many such angles, but we have agreed to choose the unique y such that $-\pi/2 \leq y \leq \pi/2$. This value is $\pi/6$, so $\arcsin 1/2 = \pi/6$.

Example 78 Find $\arcsin -\sqrt{3}/2$.

Solution: If $y = \arcsin \sqrt{3}/2$, then $\sin y = -\sqrt{3}/2$ and $-\pi/2 \leq y \leq \pi/2$. Hence $y = -\pi/3$.

Example 79 Find $\arcsin(\sin \pi/5)$.

Solution: We let $y = \arcsin(\sin \pi/5)$, and rewrite this statement as $\sin x = \sin(\pi/5)$. We know that there are many solutions to this equation: $x = \pi/5, 4\pi/5$, and so on. But since we require that $-\pi/2 \leq y \leq \pi/2$, so $\arcsin(\sin \pi/5)$ is just $\pi/5$.

Example 80 Find $\arcsin(\sin 3\pi/5)$.

Solution: As usual, we write $x = \arcsin(\sin 3\pi/5)$, so that $\sin x = \sin 3\pi/5$. But this time we cannot choose $x = 3\pi/5$, since this value is not in the required interval. However, there is a value of x in the interval that satisfies this equation. It is $x = 2\pi/5$, and this is our required value.

Example 81 Draw the graph of the function $y = \sin(\arcsin x)$.

Solution: We first decide what the domain of definition of this function is. Since we are taking $\arcsin x$, we must have $-1 \leq x \leq 1$. And since y is the sine of some angle, $-1 \leq y \leq 1$ as well. On these intervals, $\sin(\arcsin x)$ is simply x, so the graph is as follows:

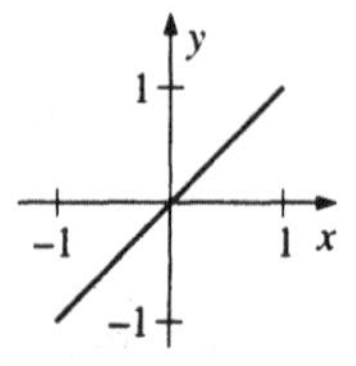

[2]We can explain the odd notation $y = \arcsin x$ by remembering that it stands for the sentence "y is the arc (or angle) whose sine is x".

Example 82 Draw the graph of the function $y = \arcsin(\sin x)$.

Solution: We will soon see that this is not the same as the previous example(!). Again, we begin by deciding on the domain of the function. We can take the sine of any real number x. Since the resulting value is in the interval from -1 to 1, we can then take the arcsine of this value. Hence the function $y = \arcsin(\sin x)$ is defined for any real number x. The possible values for y are those of the arcsine function, so $-\pi/2 \leq y \leq \pi/2$.

Let us next look at the function for values of x between $-\pi/2$ and $\pi/2$. On this interval, we find that $y = x$, so the graph looks like this:

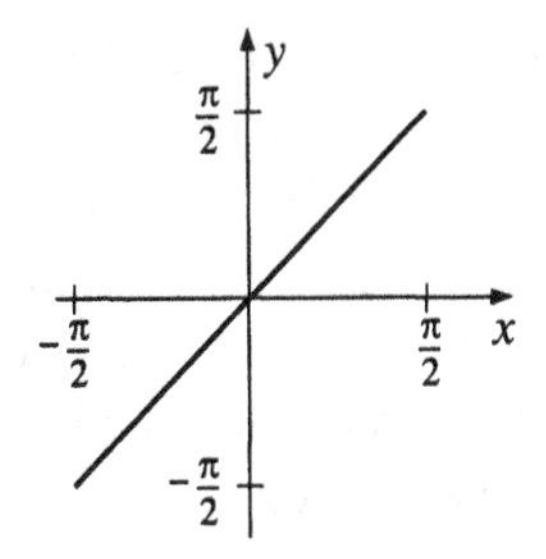

But x can take on any real value, so we are not finished. Let us look at the function for values of x between $\pi/2$ and $3\pi/2$. In this interval, $\sin x$ decreases from 1 to -1, so the values of $y = \arcsin(\sin x)$ will decrease from $\pi/2$ to $-\pi/2$. The reader is invited to check that the graph is the following:

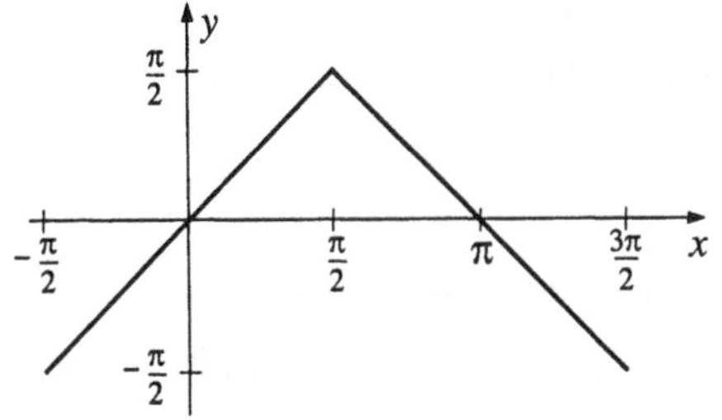

And now we note that the function $y = \arcsin(\sin x)$ is periodic, with period 2π: $\arcsin(\sin[x+2\pi]) = \arcsin(\sin x)$. The full graph is as follows:

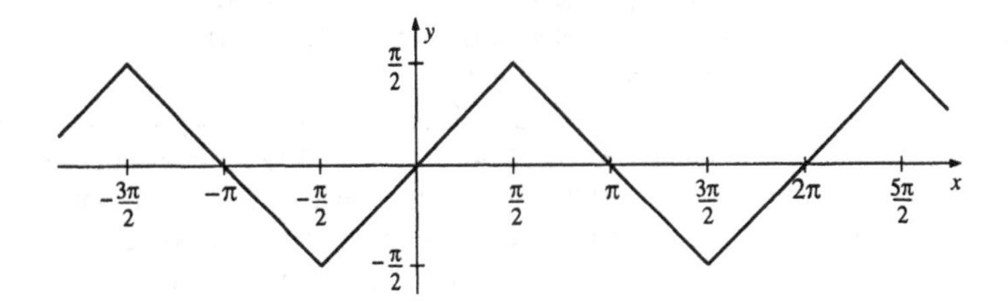

As we mentioned, the arcsine function appears on your calculator, and you will find that the calculator knows how to compute $\arcsin x$ for any number x. The button which does this is marked either [arcsin] or [$\sin^{-1}$]. (We are unhappy with the second notation, and will use only the first.[3])

In just the same way, we can define an inverse of the function $y = \tan x$. We choose an interval near 0 for which the function is monotone. It will be convenient once again to choose the interval $-\pi/2 \leq x \leq \pi/2$. Then the inverse function, which we will call arctan x, will take on all real values.

This is really a very nice function, since it is defined for any real number x. Its y-values, however, are restricted to the interval $-\pi/2 \leq x \leq \pi/2$. Indeed, the function $y = \arctan x$ supplies us with a one-to-one correspondence between all the real numbers and the numbers on that interval.[4]

We can also define an inverse of the function $y = \cos x$. But we cannot choose the same interval we chose for the sine and tangent, since the cosine is not monotone on $-\pi/2 \leq x \leq \pi/2$. Instead we choose the interval $0 \leq x \leq \pi$, on which the function $y = \cos x$ is monotone and decreasing. We write the new function $y = \arccos x$.

Example 83 Find $\sin(\arccos(5/13))$.

Solution: Let $\alpha = \arccos 5/13$. Then $\cos\alpha = 5/13$, $0 \leq \alpha \leq \pi$, and we seek $\sin\alpha$. This is a problem we've seen before. We find that $\sin\alpha = 12/13$.

Example 84 Find $\cos(\arcsin(-3/5))$.

Solution: Let $\alpha = \arcsin(-3/5)$. Then $\sin\alpha = (-3/5)$, $-\pi/2 \leq \alpha \leq \pi/2$, and we seek $\cos\alpha$. This time, α is in quadrant IV, so $\cos\alpha = 4/5$, a positive number.

In summary,

$$y = \arcsin x \text{ means } x = \sin y \text{ and } -\pi/2 \leq x \leq \pi/2$$

$$y = \arccos x \text{ means } x = \cos y \text{ and } 0 \leq x \leq \pi$$

$$y = \arctan x \text{ means } x = \tan y \text{ and } -\pi/2 \leq x \leq \pi/2.$$

---

[3]The notation $\sin^{-1} 1/3$ looks too much like the notation $\sin^2 1/3$, which of course means $(\sin 1/3)(\sin 1/3)$. By analogy, the symbol $\sin^{-1} 1/3$ "should" mean $1/\sin(1/3) = \csc 1/3$. But it means something completely different. While it remains standard in some texts, and on some calculators, we will not use it.

[4]One way to understand this is to say that there are "just as many numbers" on the whole line as there are on the interval $-\pi/2 \leq x \leq \pi/2$. When mathematicians started talking like this, some people thought this statement strange, since the interval has finite length while the line is infinite in length. What they meant, however, was simply that the notion of "length" is not based on the "number" of points in the segment being measured.

Exercises

1. Find the value of:
   (a) $\arcsin 0.5$ (b) $\arccos 0.5$ (c) $\arctan 1$
   (d) $\arcsin(-\frac{\sqrt{3}}{2})$ (e) $\arccos(-\frac{\sqrt{3}}{2})$ (f) $\arctan(-\sqrt{3})$
   (g) $\arcsin 2$

2. Find the numerical value of the following expressions:
   (a) $\sin(\arcsin 0.5)$ (b) $\cos(\arccos \frac{\sqrt{3}}{2})$ (c) $\tan(\arctan(-1))$
   (d) $\arcsin(\sin \frac{\pi}{3})$ (e) $\arccos(\cos \frac{11\pi}{6})$

3. Show that $\sin(\arccos b) = \pm\sqrt{1-b^2}$. What determines whether we should choose the positive sign or the negative sign?

4. Express $\tan(\arcsin b)$ in terms of b. Will we need an ambiguous sign, as we did in Problem 3?

5. Express $\cos(\arctan b)$ in terms of b.

6. Show that $\arccos(\sin\alpha) = \pi/2 - \alpha$, for $0 \le \alpha \le \pi/2$. What can you say for values of α outside this set?

7. Find each of the following values:
   (a) $\arcsin(\sin \frac{\pi}{11})$ (b) $\arcsin(\sin \frac{2\pi}{11})$ (c) $\arcsin(\sin \frac{3\pi}{11})$
   (d) $\arcsin(\sin \frac{4\pi}{11})$ (e) $\arcsin(\sin \frac{5\pi}{11})$ (f) $\arcsin(\sin \frac{6\pi}{11})$
   Hint: For most students, Part (f) is much more difficult than the others.

8. Draw the graph of the function $y = \cos(\arccos x)$.

9. Draw the graph of the function $y = \arccos(\cos x)$.

10. Find the numerical value of $\sin(\arcsin 3/5 + \arcsin 5/13)$. (Hint: Let $\alpha = \arcsin 3/5$, $\beta = \arcsin 5/13$, and use the formula for $\sin(\alpha+\beta)$.)

11. Recall that $\tan(\alpha+\beta) = (\tan\alpha + \tan\beta)/(1 - \tan\alpha\tan\beta)$. Using this formula, prove that $\arctan a + \arctan b = \arctan \frac{a+b}{1-ab}$.

12. The diagram below shows three equal squares, with angles α, β, γ as marked. Prove that $\alpha + \beta = \gamma$.

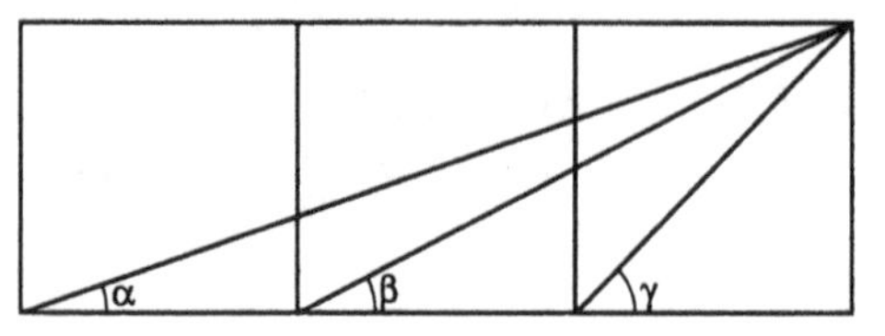

Hint: Note that $\alpha = \arctan 1/3$, $\beta = \arctan 1/2$, and $\gamma = \arctan 1$. Then use the formula from Problem 8.

13. Extra credit: Can you prove the result in Problem 9 without using trigonometry?

3 Graphing inverse functions

How is the graph of a function related to the graph of its inverse function?

Example 85 Let $y = x^2$, for $x \geq 0$ and $y \geq 0$. As we have seen, it is monotone increasing. Here is its graph:

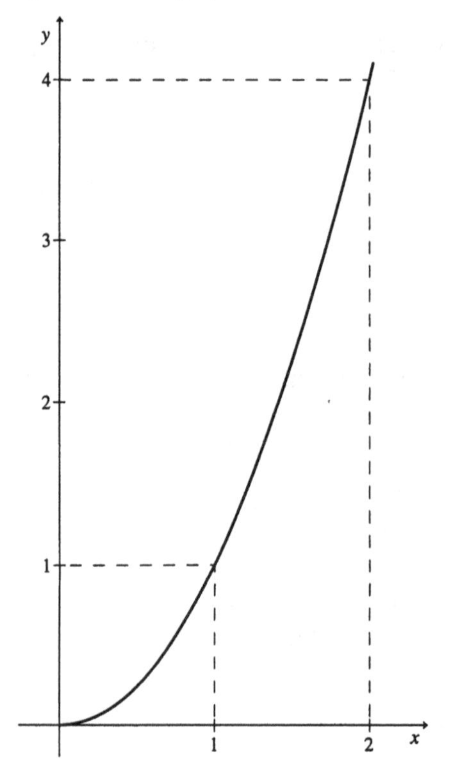

We can read the values of the function from the graph. For example, the diagram shows that $f(2) = 4$, since the x-value 2 corresponds to the y-value 4 on the graph.

The inverse function, as we have seen, is $g(y) = \sqrt{y}$. This graph also contains all the information we need to find values of the inverse function. We just choose our first number on the y-axis, and use the graph to get the corresponding number on the x-axis. For example, if we want $g(4)$, we find the number 4 on the y-axis, and use the graph to find the corresponding number (which is 2) on the x-axis.

However, many people are more comfortable using the letter x to denote the number in set A for which the function is making an assignment, and the letter y for the number in set B to which x is assigned. There are two ways to accommodate this need. We can simply relabel the axes of the original graph:

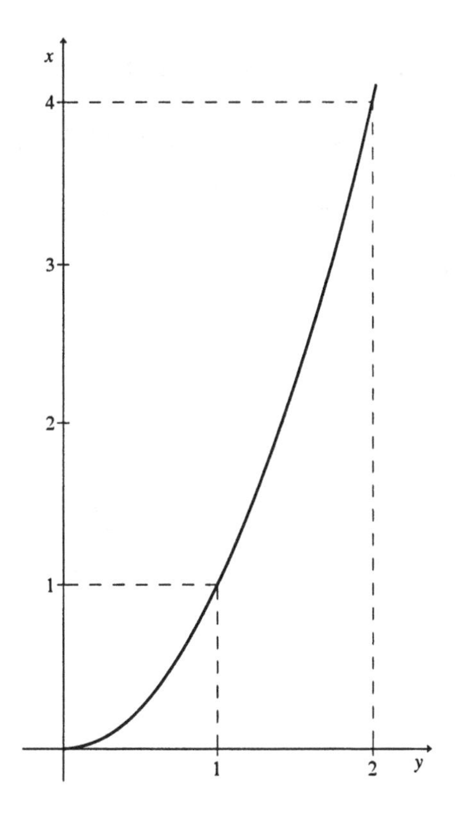

But many people prefer the x-axis to appear horizontal, and the y-axis to appear vertical, on the page. We can accommodate them by reflecting the graph around a diagonal line:

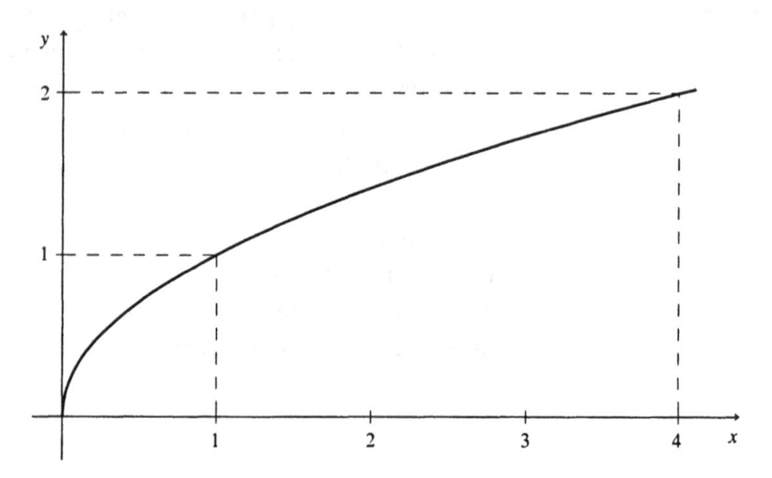

This graph contains the same information as the others, but in a more conventional form.

Here are graphs of the sine function, and its inverse, the arcsine function. The graph of the inverse function is given in the conventional position. Note that the domains are restricted as we discussed above.

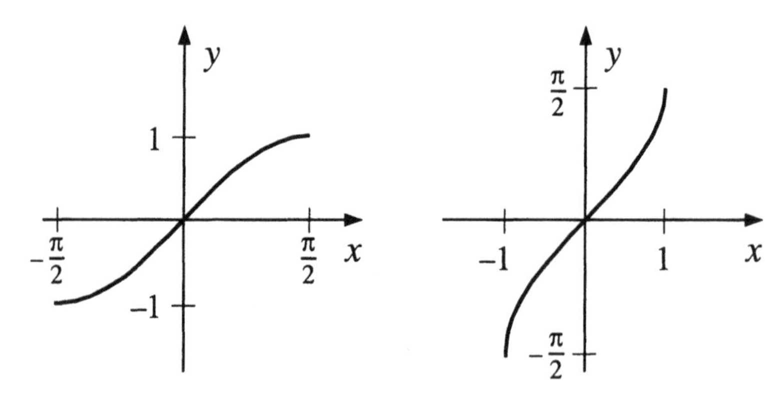

And here are graphs of $y = \arccos x$ and $y = \arctan x$:

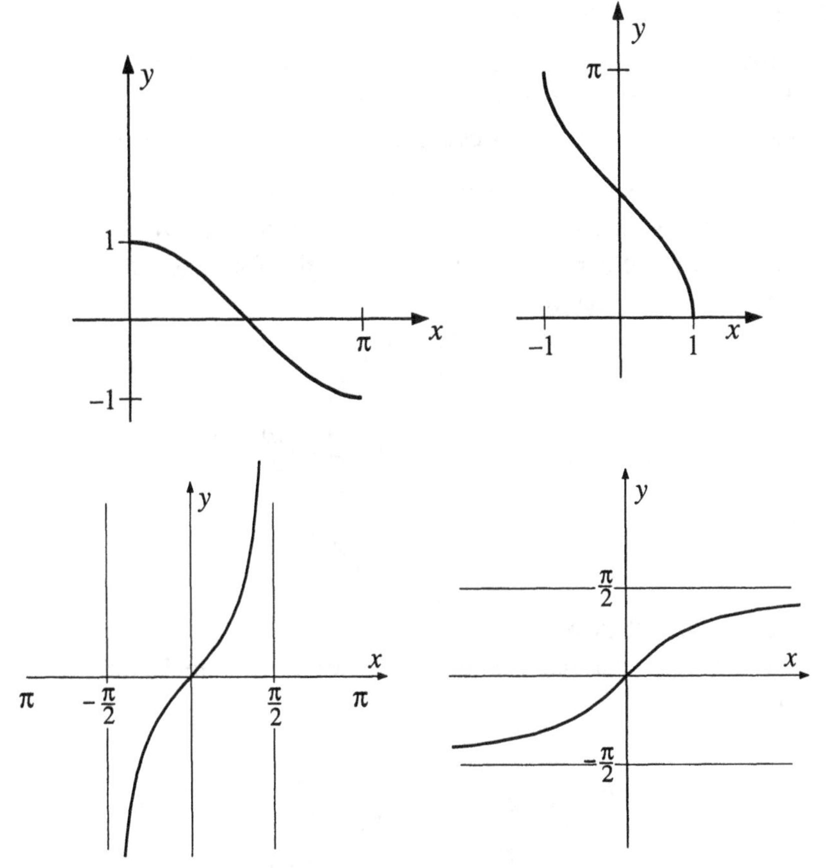

The graph of $y = \arctan x$ shows clearly how the function maps the entire real line onto a finite interval.

4 Trigonometric equations

We must often solve trigonometric equations: equations in which trigonometric functions of the unknown quantity appear. We can often use the following method to solve these:

1. Reduce them to the form $\sin a = a$, $\cos x = a$, or $\tan x = a$;
2. Locate the solutions to these simple equations between 0 and 2π;
3. Use the periods of the functions $\sin x$, $\cos x$, and $\tan x$ to find all the solutions.

We start with a simple example.

Example 86 Solve the equation $\sin x = 1/2$.

This means that we must find *all* the values of x for which $\sin x = 1/2$.

We will describe two ways of finding these values. Our first method uses a circle, and our second uses a graph of the function $y = \sin x$.

Solution 1: We first use a unit circle, centered at the origin. As a first step, we find two particular answers. We recall that $\sin \pi/6 = 1/2$. Let us illustrate this on our circle. We draw an angle of $\pi/6$, and find the line segment which is equal to $1/2$:

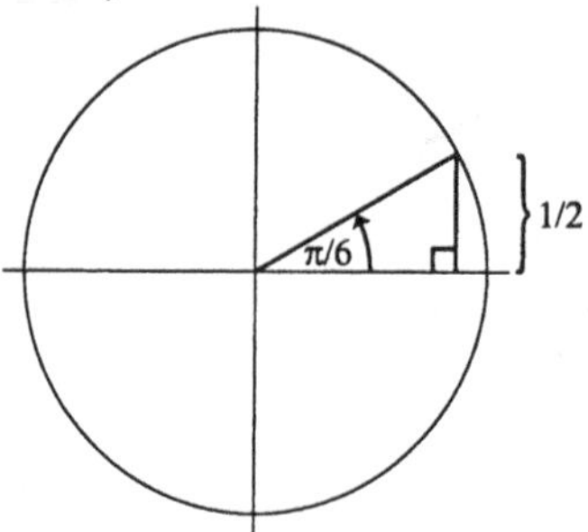

This is our first answer.

But if we draw a horizontal line across the circle, we find another angle whose sine is $1/2$:

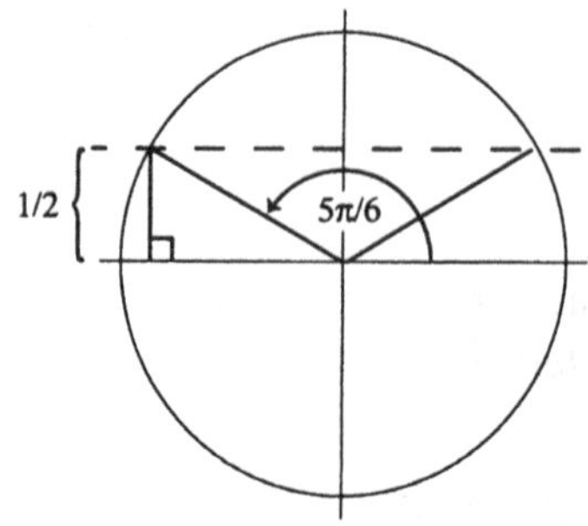

So we have the answers $x = \pi/6$ and $x = 5\pi/6$. These are all the possible answers in the interval $0 \leq x \leq 2\pi$.

To find more answers, note that we can make as many complete rotations about the circle as we like (either clockwise or counterclockwise), and we will get back to the same point.

From our first answer, we get the new values $\pi/6 \pm 2\pi$, $\pi/6 \pm 4\pi$, $\pi/6 \pm 6\pi$, and so on. We can write these as $\pi/6 + 2\pi n$ for any integer n.

From our second answer, we get the new values $5\pi/6 \pm 2\pi$, $5\pi/6 \pm 4\pi$, $5\pi/6 \pm 6\pi$, and so on. We can write these as $5\pi/6 + 2\pi n$ for any integer n.

So we have two sequences of answers:

$$\pi/6 + 2\pi n \quad \text{for any integer } n\text{, and}$$
$$5\pi/6 + 2\pi n \quad \text{for any integer } n.$$

These two sequences contain all the solutions to our equation.

We can express all these solutions more elegantly. We started with the basic answers $x = \pi/6$ and $5\pi/6$. We can write $5\pi/6$ as $\pi - \pi/6$. Then the second set of answers will be $\pi - \pi/6 + 2\pi n$. Then we can write the two sequences of solutions as:

$$2n\pi + \pi/6 \quad \text{for any integer } n\text{, and}$$
$$(2n+1)\pi - \pi/6 \quad \text{for any integer } n.$$

Now we note that the expression $2n\pi$ represents any even integer multiple of π, and we must add $\pi/6$ to this to get an answer to our equation, while $(2n+1)\pi$ represents any odd integer multiple of π, and we must subtract $\pi/6$ to get an answer. So we can write our solutions elegantly as:

$$\pi k + (-1)^k(\pi/6) \quad \text{for any integer } k.$$

The reader can verify that for $k = 2n$ (that is, for an even integer k), we obtain the first sequence of solutions, and for $k = 2n+1$ (for odd integers k), we obtain the second sequence.

Solution 2: We can use the graph $y = \sin x$ to solve our equation. Along with the graph of the function $y = \sin x$, we draw the line $y = 1/2$:

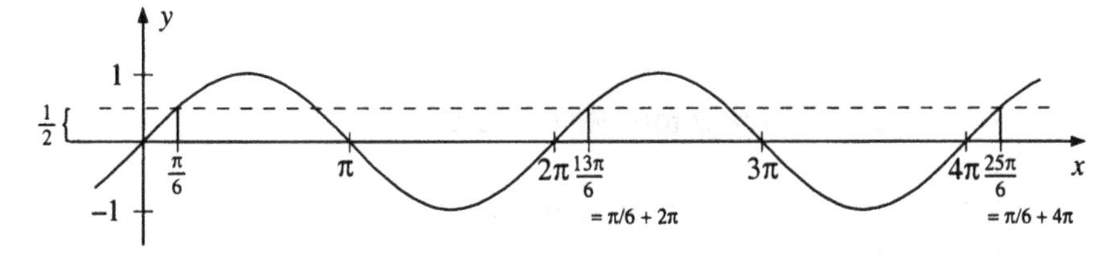

This line intersects the graph at a point whose x-coordinate is $\pi/6$. This is our first initial solution. Since the graph of $y = \sin x$ has period 2π, we will find more solutions, whose x-coordinates are $\pi/6 \pm 2\pi$, $\pi/6 \pm 4\pi$, $\pi/6 \pm 6\pi$, and so on.

The line $y = 1/2$ also intersects the graph at the point whose x-coordinate is $5\pi/6$. This is our second initial solution. Again, periodicity gives us more solutions, whose x-coordinates are $5\pi/6 \pm 2\pi$, $5\pi/6 \pm 4\pi$, $5\pi/6 \pm 6\pi$, and so on.

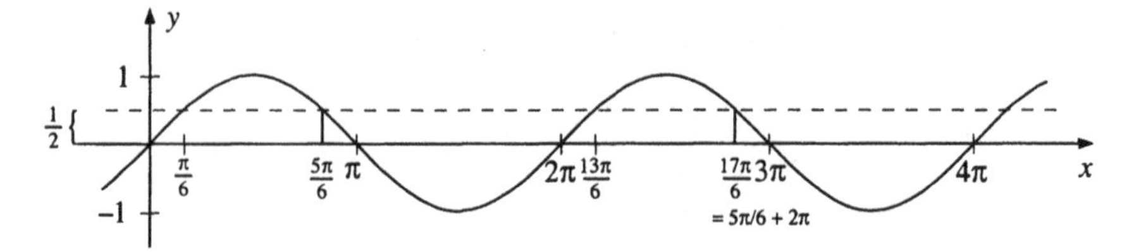

So, as before, we have two sequences of solutions: $\pi/6 + 2\pi n$ and $5\pi/6 + 2\pi n$, and we can express them elegantly as $\pi k + (-1)^k(\pi/6)$, for any integer k.

> *In general, if we need to solve a simple trigonometric equation, we can first find all the solutions between* 0 *and* 2π, *then use periodicity to get all the other solutions*

Exercises

1. Using the graph above, find all the points x on the x-axis such that $\sin x > 1/2$.
2. Solve the equation $\sin x = -1/2$.
3. Solve the equation $\cos x = \sqrt{2}/2$.
4. Solve the equation $\tan x = 1$.
5. Solve the equation $\sin x = -1$.

5 A more general trigonometric equation

Take some acute angle α. We wish to solve the equation $\sin x = \sin \alpha$. One solution is immediate: $x = \alpha$.

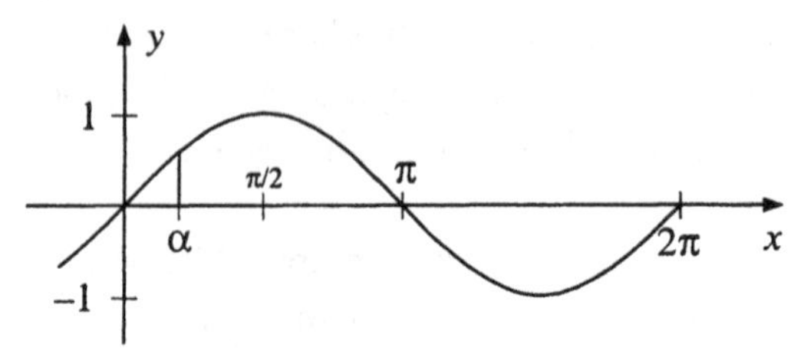

Periodicity then gives us the sequence of solutions $\alpha + 2n\pi$, for any integer n.

We also have a second immediate solution: $x = \pi - a$.

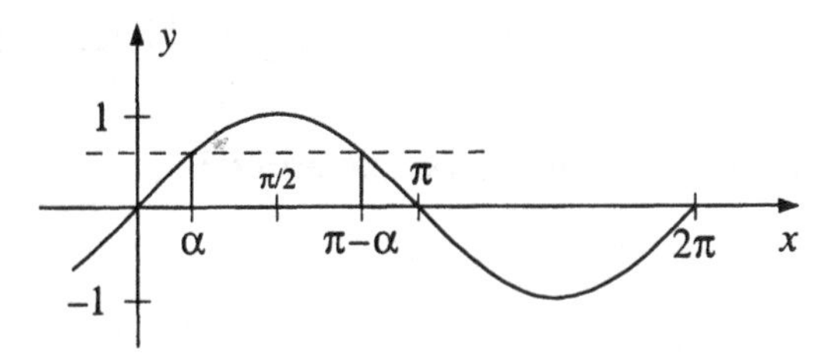

This solution gives us a second sequence of solutions, which we can write as $\pi - a + 2n\pi$, for any integer n.

So the solutions are given by

$$\alpha + 2n\pi \quad \text{and} \quad \pi - a + 2n\pi$$

for any integer n. As before, we can state this result more elegantly as

$$k\pi + (-1)^k\alpha \ .$$

Exercises

1. Solve the equation $\sin x = \sin \pi/5$.

2. Solve the equation $\sin x = \sin \pi/2$.

3. Using the graph of the function $y = \cos x$, show that the solutions to the equation $\cos x = \cos \alpha$ (for some acute angle α) are given by $2\pi n + \alpha$ and $2\pi n - \alpha$, for any integer n.

4. Solve the equation $\cos x = \cos \pi/5$.

5. Using the graph of the function $y = \tan x$, show that the solutions to the equation $\tan x = \tan \alpha$ (for some acute angle α) are given by $\alpha + \pi n$, for any integer n.

6. Solve the equation $\tan x = \tan \pi/5$.

7. Suppose α is some fixed angle. Express in terms of α all the solutions to the equation $\sin x = -\sin \alpha$. (Hint: One approach is to recall that $-\sin \alpha = \sin(-\alpha)$.)

8. Suppose α is some fixed angle. Express in terms of α all the solutions to the equation $\cos x = -\cos\alpha$.

9. Check that the formula $x = (-1)^n\alpha + \pi n$ represents all the solutions to the equation $\sin x = \sin\alpha$, as n takes on all integer values.

6 More complicated trigonometric equations

Example 87 Solve the equation $\cos^2 x = 3/4$.

Solution 1: This equation is equivalent to the two equations

$$\cos x = \frac{\sqrt{3}}{2} \text{ and } \quad \cos x = -\frac{\sqrt{3}}{2}.$$

The first equation has two solutions between 0 and 2π. They are $x = \pi/6$ and $x = 11\pi/6$:

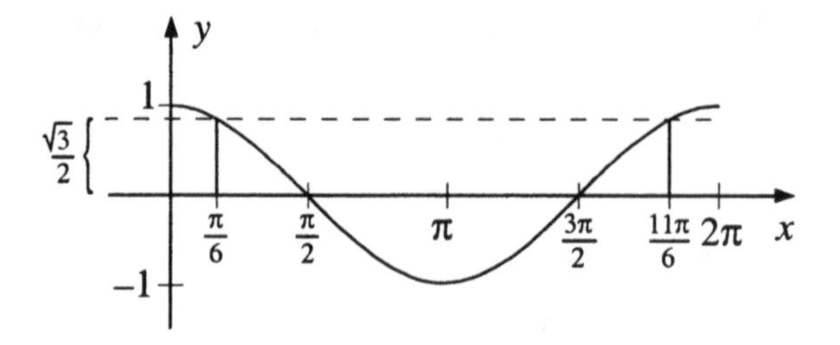

Then periodicity gives us two sequences of solutions for our first equation: $x = \pi/6 + 2\pi n$ and $x = 11\pi/6 + 2\pi n$, for any integer n.

Now we turn to our second equation. The equation $\cos x = -\sqrt{3}/2$ has two solutions between 0 and 2π, namely, $5\pi/6$ and $7\pi/6$.

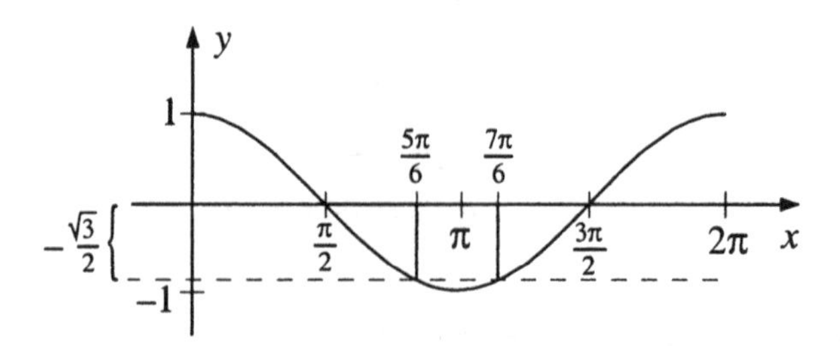

This gives two more sequences of solutions: $x = 5\pi/6 + 2\pi n$ and $x = 7\pi/6 + 2\pi n$, for any integer n.

Altogether, there are four sequences of solutions:

$$x = \pi/6 + 2\pi n \quad \text{for any integer } n$$
$$x = 5\pi/6 + 2\pi n \quad \text{for any integer } n$$
$$x = 7\pi/6 + 2\pi n \quad \text{for any integer } n\text{, and}$$
$$x = 11\pi/6 + 2\pi n \quad \text{for any integer } n.$$

The reader is invited to try to find one elegant formula that will give all these solutions.

Solution 2: [Outline] We can write $\cos 2x = 2\cos^2 x - 1 = 2(3/4) - 1 = 1/2$. Then we solve for $2x$ (for example, by looking at the graph of $y = \cos 2x$) to find the four sequences of values for $2x$. Finally, we divide each value we find by 2, to solve for x.

Example 88 Solve the equation $\sin x = \cos x$.

Solution 1: We can recall that $\cos x = \sin(\pi/2 - x)$, and rewrite the equation in terms of the sine function:

$$\sin x = \sin(\pi/2 - x) .$$

But, as we saw earlier, the equation $\sin x = \sin\alpha$ has two sequences of solutions:

(i) $x = \alpha + 2\pi n$ for any integer n, or

(ii) $x = (\pi - \alpha) + 2\pi n$ for any integer n.

We apply this result with $\alpha = \pi/2 - x$. From sequence (i) we get $x = \pi/4 + \pi n$. From sequence (ii) we get the equation $x = \pi/2 + x + 2\pi n$, which has no solutions at all.

Thus the solutions to the equation $\sin x = \cos x$ are given by the sequence

$$x = \pi/4 + \pi n \quad \text{for any integer } n.$$

Solution 2: If we divide both sides of the equation by $\cos x$, we obtain a new equation involving only the tangent function: $\tan x = 1$.

The only solution between 0 and π is $x = \pi/4$. Since the period of the tangent function is π, this initial solution give all the others, which can be written as $\pi/4 + \pi n$, for any integer n.

A fine point: If we divide by $\cos x$, we must check that this expression cannot be equal to 0. In fact, if $\cos x = 0$, we cannot have $\cos x = \sin x$, because the two functions are never 0 for the same value of x.

Example 89 Solve the equation $\sin x = \cos 2x$.

Again, we offer two solutions.

Solution 1: We rewrite the equation in terms of the sine function, and proceed as in the second solution to Example 87. We have $\sin x = \sin(\pi/2 - 2x)$.

We can now distinguish two cases, as we did in Solution 1 to Example Example 88. If $x = (\pi/2 - 2x) + 2\pi n$, then $x = \pi/6 + 2\pi n/3$, which gives one sequence of solutions.

In the second case, we have $x = (\pi - (\pi/2 - 2x)) + 2\pi n$. This leads to $x = -\pi/2 - 2\pi n$. This is a second sequence of solutions.

Solution 2: We know that $\cos 2x = 1 - 2\sin^2 x$ (see Chapter 7). So we can rewrite the given equation as

$$\sin x = 1 - 2\sin^2 x \quad \text{or} \quad 2\sin^2 x + \sin x - 1 = 0\,.$$

Let us try to solve for $\sin x$ by factoring (if this doesn't work, we can always use the quadratic formula). We have $(2\sin x - 1)(\sin x + 1) = 0$, so $\sin x = 1/2$ or $\sin x = -1$.

We can solve these equations separately, using the methods we have already demonstrated.

For $\sin x = 1/2$, we find $x = \pi/6 + 2\pi n$ or $x = 5\pi/6 + 2\pi n$, for any integer n.

For $\sin x = -1$, we have $x = 3\pi/2 + 2\pi n$, for any integer n.

There are three sequences of solutions.

Example 90 Solve the equation $\tan^2 x = 3$.

Solution: The equation is equivalent to the two equations

$$\tan x = \sqrt{3} \text{ and } \tan x = -\sqrt{3}\,.$$

An initial answer to the first equation is $x = \pi/3$, and periodicity gives the answers $\pi/3 + \pi n$, for any integer n.

The second equation has an initial solution $x = -\pi/3$, and periodicity gives the answers $-\pi/3 + \pi n$, for any integer n. These two sequences give the complete solution.

In conclusion, we note that we have already shown (Ch. 7, Appendix I.2; p. 159), that any trigonometric identity can be reduced to an algebraic identity. The same is true for trigonometric equations. However, the algebraic equation that results is often more difficult than the same equation in trigonometric form.

Exercises

1–12. Find the solution sets for the following equations:

1. $\sin 2x = 1$ 2. $\sin x/2 = 1/2$
3. $\cos x = \sin 2x$ 4. $\sin x = \sin 3x$
5. $\cos x = \sin 4x$ 6. $26\sin^2 x + \cos^2 x = 10$
7. $\cos^2 x - \cos x = \sin^2 x$ 8. $3\tan^2 x = 12$
9. $\cos 2x = 2\sin^2 x$ 10. $\tan^2 x = \cot x$
11. $\dfrac{5}{\cos^2 x} = 7\tan x + 3$ 12. $\sqrt{3}\tan^2 x + 1 = (1+\sqrt{3})\tan x$

13. Let us look back at Example 89. Solution 1 gave the general solution as

$$x = \pi/6 + 2\pi n/3 \quad \text{or}$$
$$x = -\pi/2 - 2\pi n \quad \text{for any integer } n.$$

But Solution 2 gave the general solution

$$x = \pi/6 + 2\pi n \quad \text{or}$$
$$x = 5\pi/6 + 2\pi n \quad \text{or}$$
$$x = 3\pi/2 + \pi n \quad \text{for any integer } n.$$

Show that these two sets of solutions are actually identical.

Appendix – The Miracles Revealed

In Chapter 5 we discussed two small miracles:

The Miracle of the Tangent

If we draw a tangent to the curve $y = \sin x$ at the point $x = \alpha$, then the distance between d, the point of intersection of this tangent with the x-axis, and the point $(\alpha, 0)$ is $|\tan\alpha|$.

The Miracle of the Arch

The area under one arch of the curve $y = \sin x$ is 2.

We now return to these results and furnish their proofs. Each draws on techniques that are standard in the study of the calculus. In particular, each uses the fact that the quotient $\sin h/h$ approaches 1 as h gets close to 0. We showed why this is true on Chapter 5, p. 118. A more rigorous proof would involve the notion of limit, which is the fundamental notion of the calculus. In this section, we give a sketch of a proof for each miracle that parallels the more formal approach used in a course on calculus.

Proof of The Miracle of the Tangent

The diagram shows a point $P(\alpha, \sin\alpha)$ on the curve $y = \sin x$. It intersects the x-axis at point R. We will show that $QR = |\tan\alpha|$, by writing an equation for line PR, then finding the coordinates of point R.

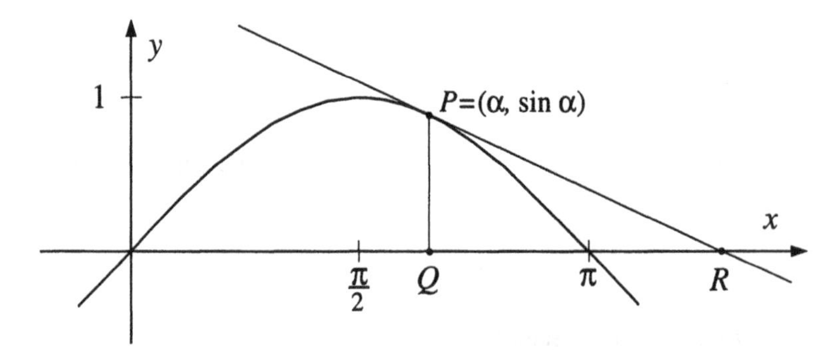

We can write the equation of a line using the coordinates of a point on the line and the line's slope. The point will be P, with coordinates $(\alpha, \sin\alpha)$.

To get the slope of line PR, we use a technique from the calculus. Instead of looking at tangent PR, we look at a secant to the curve $y = \sin x$, which intersects the curve near point P. We take two points, A and B, one just to the left of P and one just to the right, at a small distance h along the x-axis:

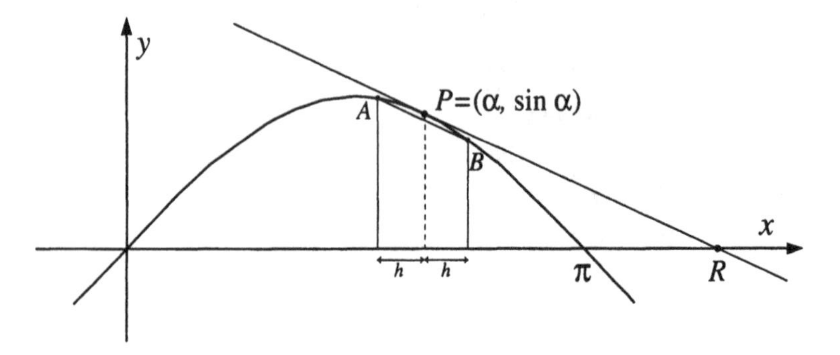

The coordinates of point A are $(\alpha - h, \sin(\alpha - h))$, and the coordinates of B are $(\alpha + h, \sin(\alpha + h))$. From these two points we can compute the slope of secant AB:

$$\begin{aligned}
&\frac{\sin(\alpha + h) - \sin(\alpha - h)}{2h} \\
&= \frac{\sin\alpha\cos h + \cos\alpha\sin h - (\sin\alpha\cos h - \cos\alpha\sin h)}{2h} \\
&= \frac{2\cos\alpha\sin h}{2h} = \cos\alpha\left(\frac{\sin h}{h}\right).
\end{aligned}$$

Now we take smaller and smaller (positive) values of h, so that points A and B get closer together, and secant AB begins to resemble tangent PR. The expression $\sin h/h$ gets closer and closer to 1 as h approaches 0. And of course $\cos\alpha$ does not change as h approaches 0. So the slope of secant AB, which is looking more and more like tangent PR, gets closer and closer to the value $\cos\alpha$. It is reasonable, then, to expect that the slope of PR is exactly $\cos\alpha$. (In calculus, this technique of finding the slope of a tangent to a curve will receive a full justification. It is related to the notion of the *derivative* of a function.)

Now we can find the equation of line PR, through point $P(\alpha, \sin\alpha)$ and with slope $\cos\alpha$:

$$\frac{y - \sin\alpha}{x - \alpha} = \cos\alpha .$$

We need the x-coordinate of point R. Its y-coordinate is 0, so its x-coordinate is obtained by letting $y = 0$ in the equation above. We find that

$$x = \alpha - \frac{\sin\alpha}{\cos\alpha} = \alpha - \tan\alpha .$$

Then the length of QR is just $|\alpha - (\alpha - \tan\alpha)| = |\tan\alpha|$.

Exercises

1. The diagrams above show a case where $\alpha > \pi/2$. Take a numerical value of α slightly larger than $\pi/2$ (for example, $\alpha = 1.6$), and follow the argument above. (Note that for such values value of α, $\tan\alpha < 0$.)

2. Take a value of α between 0 and $\pi/2$, and follow the argument again. Note that for such values of α, $\tan\alpha > 0$. Where does point R fall in these cases?

3. When does point R fall on the origin?

4. Where does point R fall when α is very close to $\pi/2$? How does your answer depend on whether α is greater or less than $\pi/2$?

Proof of the Miracle of the Arch

This miracle concerns the area under one arch of the curve $y = \sin x$, which we claim is exactly 2. On p. 117 we showed that this area A satisfies the inequalities $\pi/2 < A < \pi$. We did this by drawing figures bounded by straight lines that approximated the area A. We can improve on this approximation by taking regions closer and closer to the region whose area we want to measure. We will construct these regions out of rectangles.

We take the interval from 0 to π along the x-axis, and divide it into many equal pieces. If there are n of these pieces, then the points of division are $x_0 = 0$, $x_1 = \pi/n$, $x_2 = 2\pi/n$, $\ldots x_{n-1} = (n-1)\pi/n$, and $x_n = n\pi/n = \pi$. For each point x_i, we draw a rectangle by erecting a perpendicular to the x-axis with one endpoint at x_i and the other on the curve $y = \sin x$ (the diagram shows the case $n = 8$):

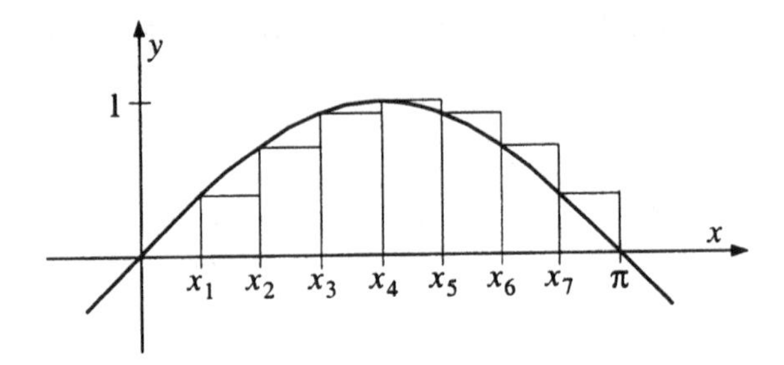

Note that the rectangles are inscribed in the arch for $0 < x_i < \pi/2$, and they are circumscribed for $\pi/2 < x < \pi$. Also, the widths of the rectangles are all π/n. Let us set $h = \pi/n$. Then as h approaches 0, the rectangles get thinner and more numerous, and the sum of their areas approaches the area A.

Finally, note that the rectangles associated with $x_0 = 0$ and $x_n = \pi$ are "degenerate": their area is 0 (no matter what value we choose for n). It will be convenient for us to ignore the rectangle associated with x_0, but to include the one associated with x_n. Then we can write the sum of the areas of these rectangles as

$$h \sin x_1 + h \sin x_2 + h \sin x_3 + \cdots + h \sin x_n$$
$$= h\left(\sin \frac{\pi}{n} + \sin \frac{2\pi}{n} + \sin \frac{3\pi}{n} + \cdots + \sin \frac{n\pi}{n}\right)$$

$$= h \frac{\sin \frac{n+1}{2}\frac{\pi}{n} \sin \frac{n}{2}\frac{\pi}{n}}{\sin \frac{\pi}{2n}} = h \frac{\sin \frac{n+1}{2}\frac{\pi}{n}}{\sin \frac{\pi}{2n}} \cdot \sin \frac{\pi}{2} = h \frac{\sin \frac{n+1}{n}\frac{\pi}{2}}{\sin \frac{\pi}{2n}} .$$

If n is very large, our set of rectangles will look more and more like the area A. But as n gets very large, the fraction $(n+1)/n$ approaches the value 1. Hence our expression for A gets close to

$$h \frac{\sin \pi/2}{\sin \pi/2n} = \frac{h}{\sin h/2} .$$

Now if we let $k = h/2$ this expression is equal to $2k/\sin k = 2(k/\sin k)$. As h gets close to 0, so does k, and so the expression approaches 1. Its reciprocal, which is $k/\sin k$, also approaches 1. This means that the sum that approximates A gets close to $2 \cdot 1 = 2$, a miraculous result.

In calculus, this technique for finding the area under a curve is related to the *integral* of a function.

Exercises

1. Using a calculator, find the approximations to A given by taking $n = 4$ and $n = 8$.

2. What do you think the area under the curve $y = \sin x$ is from $x = 0$ to $x = \pi/2$?

3. Try using the method outlined above to find the area under the curve $y = \sin x$ from $x = 0$ to $x = \pi/2$. Is the result what you might have expected?

CPSIA information can be obtained
at www.ICGtesting.com
Printed in the USA
LVHW011806150621
690291LV00016B/1321